An introduction to marine life

A Museum Victoria
Field Guide to Marine Life

An introduction to marine life

Robin Wilson
Mark Norman
Anna Syme

Series editors
Gary C. B. Poore
Mark D. Norman

Other titles in the series
An introduction to marine life
Crabs, hermit crabs and allies
Shrimps, prawns and lobsters

First published by
Museum Victoria 2007
Reprinted 2010

Museum Victoria
Publishing
GPO Box 666
Melbourne Vic 3001 Australia
Tel + 61 3 8341 7370 or 8341 7536
Fax + 61 3 8341 7573
publishing@museum.vic.gov.au

PRINTED BY
BPA Print Group
DESIGN BY
Propellant

National Library of Australia
Cataloguing-in-Publication data:
Wilson, Robin, 1956-.
An introduction to marine life.

Bibliography.
Includes index.
ISBN 9780975837054 (pbk.).

1. Marine organisms - Australia.
I. Norman, Mark (Mark Douglas).
II. Syme, Anna. III. Museum
Victoria. IV. Title.

578.77

FRONT COVER IMAGES
Ornate Cowfish, *Aracana ornata* (Ostraciidae) and the Gorgonian fan coral (Melithaeidae)

BACK COVER IMAGES
Tosia neossia (Goniasteridae)
Chromodoris tasmaniensis (Chromodorididae), *Ralpharia magnifica* (Tubulariidae),
Anoplodactylus evansi (Phoxichiliidae)

PORT PHILLIP AND
WESTERNPORT
CATCHMENT AUTHORITY

Funding for the Museum Victoria Guides to Marine Life, of which this is the first, is gratefully acknowledged from the Australian Government through a grant from its Natural Heritage Trust. The grant was facilitated by the Port Phillip and Westernport Catchment Management Authority. Publication of the series is funded in part by the Victorian Coastal Council and Parks Victoria.

MUSEUM VICTORIA FIELD GUIDE SERIES TO MARINE LIFE

These field guides to marine life enable the amateur naturalist, beachcomber or environmental scientist to identify the marine animals most commonly found along the shore or in shallow waters along the coast in the state of Victoria, Australia. Southeastern Australia is characterised by a rich fauna, with many species found nowhere else. Many of those found along the Victorian coast also occur in Tasmania and southern New South Wales. Others extend along the southern coast of the continent, through South Australia and some as far as southern Western Australia. Only a small fraction is found in tropical Australia.

This series aims to cover the common animals and each book deals with a different group of animals. More species live on the Victorian shore and shallow waters than are included in each book and many more inhabit the deeper waters of Bass Strait and beyond. Information not covered is included in the papers and books referred to at the end of this book (see Further Information) and in Museum Victoria collections, research laboratories and the library. Other information can also be obtained by contacting the Discovery Centre at Museum Victoria: *discoverycentre@museumvictoria.com.au*

This guide is accompanied by supplementary information on the Museum Victoria website *http://researchdata.museum.vic.gov.au/marine*

Museum Victoria encourages individuals to explore the diversity of coastal habitats but discourages unnecessary removal of specimens from their natural environment. Museum scientists are interested in new discoveries and unusual findings can be reported to the Discovery Centre or to Reef Watch Victoria: *info@reefwatchvic.asn.au*

Reef Watch Victoria is a community-based marine monitoring program for Victoria's temperate marine environment. Divers and snorkellers conduct regular surveys at their favourite Victorian reef sites using the Reef Watch monitoring kit. The kit includes waterproof photographic charts of 150 temperate marine species, survey sheets, manual, dive slate and pencil. Surveys are validated and entered into a centralised database for use by conservation groups, scientists, coastal and marine managers, and government organisations.

CONTENTS

CONTENTS

INTRODUCTION

For many people, the first experience of marine environments is amazement at the bewildering variety of life in the oceans. Is that white growth a coral? Is it an animal or a plant? What is the difference between a shrimp and a prawn? These and many other common questions reveal our lack of familiarity with the seas. Sea anemones and corals, sea stars and sea urchins, octopuses and squids are just a few marine creatures that we never encounter on land or in freshwater. Many other creatures are even less familiar, and it is often difficult for those interested in marine life to learn more about them. The aim of this book is to introduce the diversity of life in the seas and to help newcomers to marine biology recognise the main kinds of marine organisms. The examples we have chosen focus on Victoria and southern Australia. The emphasis is on animals and plants that are commonly seen by divers, snorkellers, beachcombers, rock poolers, and by anyone with an interest in marine life.

Rockpools at Flinders in Victoria

Marine environments

To a human standing on the shore, the sea appears to be a vast, uniform habitat. However, once we look more closely we realise this is far from the truth. On land we know that rainforest, alpine meadows and deserts each have their own animals and plants. The same applies under the sea. If we don mask and snorkel we also find different habitats: seagrass meadows, sand plains and rocky reefs covered with seaweeds. In deeper, darker waters, or in caves, there are communities dominated by sponges and coral-like animals. Estuaries have mudflats rich in burrowing animals that consume buried nutrients, while on the wave-scoured ocean coast much of the marine life is firmly attached to the rocks or is constantly on the move. Each of these environments harbours a distinctive community of animals and plants adapted to local conditions.

Uniquely Australian

Much research has been conducted in recent decades into the identity and relationships of our marine life. As a result, we now know that the vast majority of marine organisms found in southern Australia occur nowhere else—they are endemic to our waters. This is due to the long isolation of southern Australia's marine fauna. Australia's tropical seas extend into the tropical Indian and Pacific Oceans, where there is an extensive patchwork of islands and reefs. However, the creatures adapted to cool seas of our southern shores cannot live in the tropics. Most species cannot cross the vast southern oceans, nor can the animals and plants of our shallow southern shores exist on the floor of the dark, cold deep sea. Australia's southern marine life, therefore, is very much isolated, and has been for many millions of years—long enough for the evolution of new species, unique to our coast.

Common Seadragon,
Phyllopteryx taeniolatus
(Syngnathidae)

Our shallow-water marine fauna is not only uniquely Australian, it is also unusually rich in species. More than 10,000 different species of marine invertebrates and plants can be found in southern Australia. This is far more than

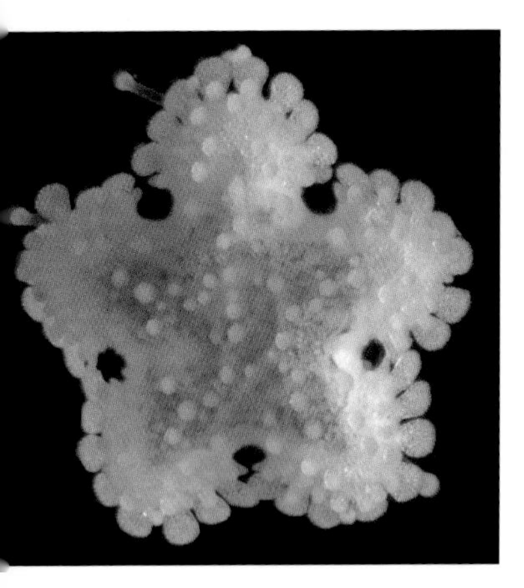

Allostichaster palmula (Asteriidae) World's smallest sea star and new species described in 2007.

would be found in a similar area of Europe or North America. If we were to include deep-water marine life in our search, that number would grow much more. New species are constantly being discovered. For example, in 2007 researchers associated with Museum Victoria discovered a new species of sea star—the smallest in the world—in Port Phillip Bay. Many other marine species have yet to be formally described and given scientific names. This vast and still poorly-known diversity means that for some kinds of marine life, it is not yet possible to find any sort of identification guide that will allow all species to be identified. The task is too great for the few taxonomists now working in Australia, and there is still much to be done by future generations of marine scientists. Fortunately, a number of community groups and amateur naturalists also have an active interest in the marine environment and many who started out as 'amateurs' are now world authorities on their area of interest. Clubs and societies interested in marine life often collaborate with scientists in museums, herbaria and other institutions, where most taxonomists are employed. The Further Information section at the end of this book will provide initial contacts for those who would like to expand their interest in marine biodiversity.

Marine conservation

Although the oceans seem vast, they are not remote from human influence. In Australia, as in most countries, humans interfere with marine life by polluting the water, by building breakwaters and marinas, by taking fish and shellfish, and in many other ways. Those areas of shoreline which remain relatively untouched will inevitably become more rare and more precious, and it is appropriate that a system of marine parks and reserves is now being established throughout Australia.

Anyone can have an impact on marine life, whether turning over rocks in a rock pool, or collecting bait or seafood. Simply walking across a tide flat or rock platform can crush some creatures. Turning rocks runs the risk of damaging

Marine national parks and other forms of marine protected areas are gaining support.

or killing some creatures, even if the rock is carefully returned to the original position. In some places, local regulations may prohibit some or all activities and your local department of fisheries or environment will provide details. Even if it is legal to collect specimens, we cannot condone doing so unless there is a good reason. Usually this will involve making prior arrangements for identification by an expert and deposition in a museum collection where the information will contribute to our broader knowledge.

Fortunately, much can be learned by low impact observation and photography in the field. Many larger species of marine life can be identified from a good photograph. Using digital cameras, consumable expenses are negligible and digital photographs can easily be numerous and detailed enough to monitor abundances, habitat preferences and seasonal movement in a variety of marine life. Becoming involved with a club or society with an interest in marine life is a good way to develop projects such as these.

Classification and names

Scientific names and classifications often seem like unnecessary jargon. However, they have two vital functions. Scientific names, because of the rules which govern formal descriptions of new species, provide a 'label' that allows us to talk about an organism without ambiguity. There are no such rules for common names: the same name can be used for different species over a wide range, or different names can be used for the same species in different regions. Thus, the species of fish known to scientists only as *Notolabrus tetricus* is variously called a rockie, rockfish, parrotfish or bluethroat wrasse. *Notolabrus* is the genus name, which used by itself refers to all the related species now placed in that genus. The other part of a scientific name is the species name—*tetricus*—but this is meaningless unless used in combination with the genus: *Notolabrus tetricus* (sometimes abbreviated as *N. tetricus*).

The second function of scientific names is to show that a species is closely related to another by classifying them together under a single group name. Thus, another Australian species of wrasse similar to our *Notolabrus tetricus* is *Notolabrus gymnogenis*. Both species are placed in the genus *Notolabrus* to show that they are thought to be closely related. Knowing about the close relatives of a species is often useful. In our example, at least four species of *Notolabrus* occur in Australia, and two more occur elsewhere–a handy starting point for the study of the evolutionary origins of biological diversity in this group of fishes. A taxonomic classification, using genus and other names is shorthand for a lot of information about evolution and relationships. This is why the researchers who name new species (taxonomists) labour to classify and name species accurately—they are not 'just names'.

Bluethroat Wrasse *Notolabrus tetricus* (Labridae)

Throughout the remainder of this book we refer to other levels (ranks) used for classifying organisms. They are the steps of classification starting from the broadest groups down to a particular species. The sequence of this hierarchy is: Phylum (plural: Phyla); Class; Order; Family; *Genus* (always shown in italics font, starting with a capital letter); *species* (always shown in italics font, starting with a lower case letter).

The most useful rank to mark the fundamental differences between living things is Phylum. Botanists use Division in place of Phylum, however since the two are equivalent (and because many small life forms are neither animal nor plant) we use Phylum for consistency. Many commonly-distinguished broad groups belong to different Phyla; for example, molluscs (clams, snails, squid and relatives) belong to Phylum Mollusca, while jellyfish, corals and sea anemones belong to Phylum Cnidaria. If we follow the classification of our species (*Homo sapiens*) we belong in the following groups:

- Phylum Chordata (animals with a rigid dorsal rod)
- Subphylum Vertebrata (animals with bony or cartilaginous vertebra)
- Class Mammalia (animals that grow hair or fur and make milk, such as humans, mice, cows, whales, bats and kangaroos)
- Order Primata (humans, apes, monkeys, marmosets, baboons and other primates)
- Family Hominidae (humans and their ancient extinct ancestors)
- Genus *Homo* (humans and their more recent extinct ancestors)
- Species *sapiens* (our single human species)

The broadest level of classification is known as Kingdom. Traditionally, plants were Kingdom Plantae and animals Kingdom Animalia. Unfortunately, life is not that simple and creatures such as fungi, bacteria, algae and many single-celled organisms are placed in different Kingdoms in different text books. There are many living things (especially in the sea) that are not clearly animal or plant and the relationships between Phyla are still uncertain. Until our understanding of these evolutionary relationships is settled, we avoid further mention of Kingdoms and commence this guide at the level of Phylum.

Scientific terminology

Due to the high diversity in form of living organisms, many special terms have been coined for the various body parts and even the sides and directions of orientation (e.g. which way is up?) These terms allow us to describe parts of organisms as concisely and accurately as possible. When first learning to identify creatures the sheer number of these terms can be daunting. The glossary at the back of this book defines the many such terms used in this guide.

'Invertebrate' is a term that is used widely but loosely in books such as this. An invertebrate is simply an animal that lacks a backbone (a column of vertebrae). Thus, the name refers not to a feature that is present, but to something that is missing, and invertebrates represent many separate branches of the evolutionary tree of life, not a single branch. Invertebrates encompass a vast variety of animals comprising many different body forms. The majority of organisms found in the sea, and in this book, are invertebrates.

Using this book

This guide is divided into two major sections. The first is a series of nine Quick Guides. The first four Quick Guides compare and contrast types of creatures that often puzzle newcomers to identifying marine life. The next two Quick Guides treat groups of commonly-encountered marine life. The last three Quick Guides treat potentially dangerous and introduced species. These pages can be used to help find the scientific name and classification for many common creatures. The Quick Guides are:

The second major section of this guide is the Gallery of Marine Life, in which all commonly encountered marine creatures are treated in their correct classification. The Gallery of Marine Life covers all creatures that are already mentioned in the Quick Guides, but now related organisms are grouped together. Each Phylum, Subphylum or Class is described and illustrated with local examples. Identification tips and comments on diversity and ecology are also provided. Those who wish to identify marine invertebrates to a

greater level of detail are directed to a Further Information section at the end of the book, where pointers are provided to relevant publications and other resources.

Becoming a trained observer

Even with books such as this, it can be difficult to know where to start when confronted by a strange marine creature. The secret is to train yourself to notice some useful feature of the mysterious animal (or is it a plant?!). Ask yourself:

- Does it have legs?
- Does it have an obvious head?
- Is it comprised of many repeated segments (or is it smooth and unsegmented)?
- Does it have a hard shell (or is it soft and squashy)?

Then, as you browse for similar organisms in these pages and illustrations, ask the same questions. When you find a possible match, use the Quick Guides to see if there is some other creature that is similar and is easily confused. Do the identification tips on the relevant Gallery of Marine Life page also agree with your specimen? Compare notes with friends or other students. Join a club such as the Field Naturalists Club of Victoria, and ask other members for help and tips. In a short while you will be observing marine life in a different, more rewarding way, and finding a new appreciation of just how many branches of the tree of life are represented right on our doorstep.

Quick Guide 1

Plants and plant-like animals

Animal or plant? Some marine animals and plants look superficially similar but are actually only distantly related. Also look at Quick Guide 2—Corals and encrusting growths.

1. Just above high water mark on rocky shores are crusts that look like flaking paint. They are neither animal nor plant but a lichen, a symbiotic partnership between a fungus and tiny algal cells (Phylum Ascomycota, p29). **2.** This furry dark green growth is common in the upper intertidal region on rocky shores throughout southern Australia. It is a blue green alga *Symploca* sp. (Phlyum Cyanophyta, p32). **3.** This shiny green-black seaweed growth is a blue-green alga *Rivularia* sp. (Phlyum Cyanophyta, p32). It grows in the intertidal zone on ocean coasts. **4.** These spongy brown growths are a brown seaweed *Colpomenia* sp. (Phlyum Phaeophyta, p38). It is commonly found on ocean coasts. **5.** Seagrasses (Phylum Anthophyta, p40) are true flowering plants. They have true leaves, stems and anchoring roots. This image of *Halophila ovalis* shows some of the tubular vessels in the leaves. **6.** *Amphibolis antarctica* is another seagrass, rather more plant-like with long stems and a dense root mass. **7, 8.** Green seaweeds like these *Caulerpa* spp. are algae (Phylum Chlorophyta, p34) not flowering plants. Seaweeds are more simple in structure than seagrasses—seaweeds have no tubular vessels in fronds and stalks. **9.** These feathery fronds are a colonial animal—a hydroid (Phylum Cnidaria, Class Hydroida, p46). Hydroids are related to corals and have tiny individual polyps with a ring of stinging tentacles. Hydroid colonies are flexible—there is no hard calcite skeleton and they do not 'crunch'.

10. Feather stars or crinoids have feathery—or ferny—looking arms that often protrude from a ledge and look a bit like brown seaweed or a colonial animal such as a bryozoan. But this is a single mobile animal with five jointed arms arranged in a star-pattern (Phylum Echinodermata, Class Crinoidea, p92). **11.** Some red seaweeds (Phylum Rhodophyta, p36) are hard and coral-like. Unlike similar-looking colonial animals such as bryozoans, seaweed fronds lack small feeding openings and are simple in structure. **12.** This small bush is not a plant, but a colonial animal—a bryozoan or lace coral (Phylum Bryozoa, p.102). The bush is made up of many separate segments, each with a single minute opening. Bryozoans have no stinging cells. Their calcite skeleton is brittle and 'crunch' if squashed hard. **13.** Sea pens are a kind of coral (Phylum Cnidaria, Class Anthozoa, p48) with small modified polyps arranged on fleshy leaf-like plates. **14.** This flower-like creature is a burrowing sea anemone (Phylum Cnidaria, Class Anthozoa, p48). Its tentacles have stinging cells to capture floating planktonic prey. **15.** Another flower-like creature. The 'flower' is the ring of radioles (feathery branches) of a polychaete worm (Phylum Annelida, Class Polychaeta, p58) in the family Sabellidae. The remainder of the worm is not visible—and if disturbed even the tentacles retract swiftly into a hidden burrow. **16.** Looking like a little feather-duster, this is the crown of feeding radioles of another polychaete worm (Phylum Annelida, Class Polychaeta, p59) in the family Sabellidae. The body of the worm always remains hidden in the tube, with only the feather-like feeding structures protruding.
17. Sometimes called sea tulip, this stalked animal is a sea squirt (Phylum Chordata, Class Ascidiacea, p108). It is only found subtidally.
18. Another kind of sea squirt (Phylum Chordata, Class Ascidiacea, p108). This is a colonial species, the bulbous tip of the stalk being made up of rows of individuals.

Quick Guide 2

Organisms that make coral-like growths

Creatures that cover rocks and pier piles with a crust-like growth are often a challenge to identify, even for experienced marine naturalists. This page presents an overview of the animals and plants that create these carpet-like colonies.

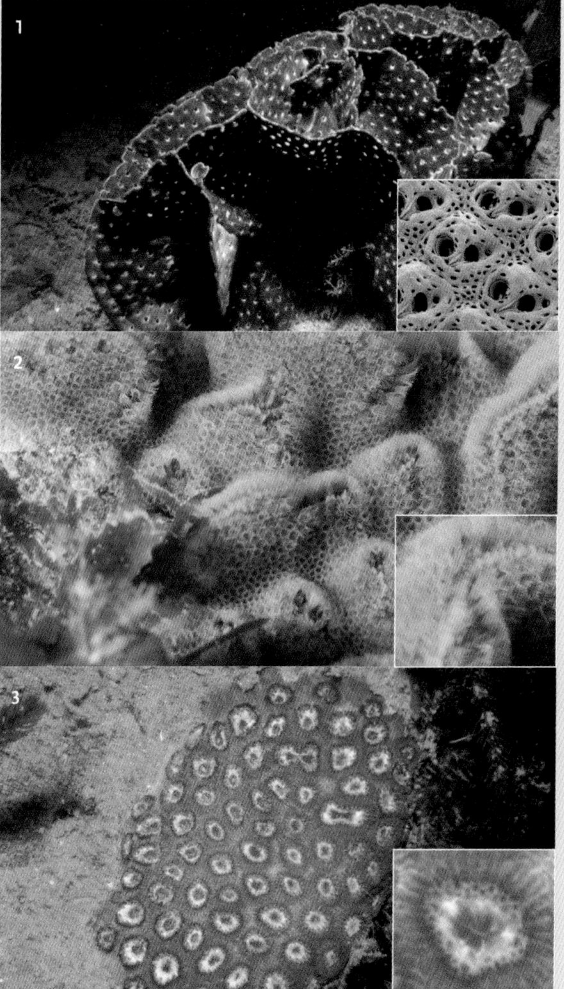

1. Lace corals or 'bryozoans' (Phylum Bryozoa, p102) are very common in southern Australia. Not all species have the lacy holes found in this species (*Adeona cellulosa*). Bryozoans may form stony colonies, but the tiny individual 'zooids' are much smaller and very different from those of true corals. The inset is a scanning electron micrograph and each zooid is about a millimetre in diameter.

2. The tiny individual 'zooids' of lace corals (Phylum Bryozoa, p102) extend minute horseshoe-shaped feathery tentacles when feeding (see inset).

3. A few hard corals (such as *Plesiastrea versipora*, illustrated) occur in southern waters (Phylum Cnidaria, Class Anthozoa, p48). They form a hard stony colony on rocks and the large polyps have a radial structure (see inset).

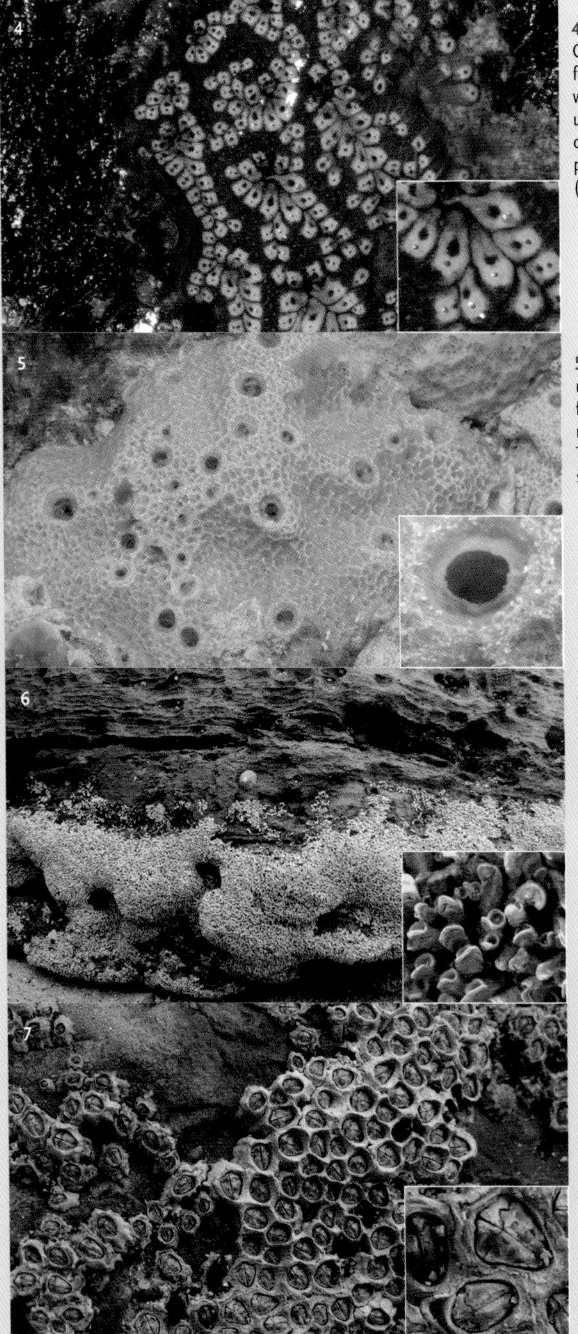

4. Compound ascidians (Phylum Chordata, Class Ascidiacea, p108) form soft encrusting growths, often with a colourful tapestry of repeated units. The colony comprises many cloned individuals, each with two pores through which water circulates (see inset).

5. Sponges (Phylum Porifera, p42) range from cushion-like to cup- and finger-shaped growths. True to their name, most feel soft and spongy. Their pores are usually irregular in size and not paired (see inset).

6. The thick white coral-like growth that is common intertidally throughout southern Australia is made by a tube worm, *Galeolaria caespitosa* (Phylum Annelida, Class Polychaeta, p58). Look closely to see the individual tubes, each closed by a tiny spiky trapdoor (see inset).

7. The Honeycomb Barnacle, *Chamaesipho tasmanica*, also forms extensive white colonies, especially on exposed intertidal shores. They can be distinguished from colonies of tube worms by their squat size and four-piece smooth trapdoor (see inset). Barnacles (Subphylum Crustacea, Class Cirripedia, p78) are highly specialised crustaceans.

Quick Guide 3

Worms, slugs and similar animals

Worms, slugs and other similar creatures represent a wide variety of only distantly related kinds of animals.

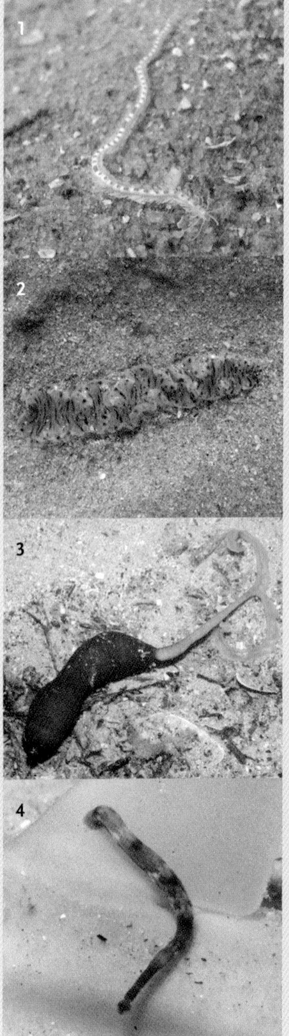

1. The most common marine worms are the polychaetes (Phylum Annelida, Class Polychaeta, p58) with regular segments along the body, each segment with paired bundles of fine bristles. **2.** Echiurans hide their bodies in the sand or a crevice; the long feeding tongue (proboscis) is normally all that is seen. This unidentified echiuran is so far only known from the proboscis—the body of the worm has not yet been seen. **3.** This echiuran, *Metabonellia haswelli,* is often shallowly hidden in the sand or a crevice but the dark green bulbous body is sometimes seen by divers who turn a few stones. Once thought to belong in their own phylum, recent studies have shown they are a type of polychaete worm (Phylum Annelida, Class Polychaeta, p58). **4.** Most leeches are terrestrial, but they also occur in the sea. Leeches are related to polychaete worms and are also segmented—they belong to the Class Hirudinea within Phylum Annelida (p58). Their most distinctive feature is the sucker at each end of the body. **5.** Peanut worms (Phylum Sipuncula, p57) are yet another type of unsegmented marine worm. They are usually tapered at the mouth end. Several species are commonly seen and all have a bumpy leathery texture. **6.** *Lepidonotus melanogrammus* is another polychaete (Phylum Annelida, Class Polychaeta, p58). It is a scale worm, so called because of the scales in two rows along the back. **7.** The sea mouse, *Aphrodita* sp., hardly looks worm-like, yet it too is a polychaete worm. The segments are easily seen if it is turned upside down (use gloves). **8.** Acorn worms (Phylum Hemichordata, p105)

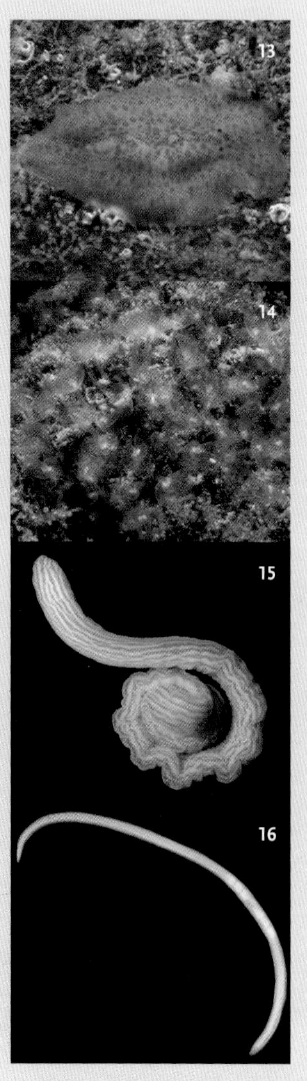

are occasionally found in shallow sediments. The body is divided into distinct regions, including the vaguely acorn-shaped proboscis, followed by a short collar. **9.** Chitons are a distinctive kind of mollusc (Order Polyplacophora, p82) characterised by a row of eight interlocking shell plates.

10. *Onchidella nigricans* is common on oceanic rocky shores. It is an air-breathing slug (Phylum Mollusca, Class Gastropoda, p84) that is slow-moving and slightly warty in texture. The underside has the muscular foot typical of gastropod snails. **11.** Sea hares (*Aplysia* species) are intertidal slugs. They sometimes reach 200 mm or more in length and discharge a purple ink when disturbed. Like other snails (Phylum Mollusca, Class Gastropoda, p84) they have a muscular foot on the underside. **12.** Some sea cucumbers, particularly the many small species, look distinctly worm-like. However, they are echinoderms (Phylum Echinodermata, Class Holothuroidea, p100) relatives of sea stars and sea urchins. Holothuroids lack segments and their skin has tiny tubular feet and/or ossicles. They often feel 'sticky'. **13.** Flatworms or polyclads (Phylum Platyhelminthes, p55) are a distinctive group of worms that are extremely flattened. These fast-moving unsegmented worms are often brownish or pale, but some are more conspicuous with brighter colour patterns. **14.** Horseshoe worms (Phylum Phoronida, p60) live in tubes—the only visible portion is a pair of horseshoe—or spiral-shaped feeding tentacles. Phoronids are widespread but not common. **15.** Ribbon worms or nemerteans (Phylum Nemertea, p56) are quite common in a variety of shallow marine habitats in southern Australia. These unsegmented worms are smooth, fragile and often have distinctive spotted or striped colour patterns. **16.** Roundworms (Phylum Nematoda, p61) include a myriad of tiny, shiny, unsegmented worms. They are usually only seen with the aid of a microscope.

Quick Guide 4

Jellyfish and other floating animals

Not all jelly-like animals are true jellyfish. These two pages are a guide to the different creatures that cruise the open ocean. Many are only seen when they wash ashore.

1. The Blue Blubber, *Catostylus mosaicus*, is a typical jellyfish (Phylum Cnidaria, Class Scyphozoa, p50). It is common in southern Australia. Like all cnidarians, it has stinging cells but in this species are usually too weak to seriously sting humans. **2.** The Lion's Mane Jellyfish, *Cyanea capillata* (Phylum Cnidaria, Class Scyphozoa, p50) can cause a painful sting and rash from a mass of trailing thin tentacles. **3.** The Moon Jelly, *Aurelia aurita*, has a four-leaf clover pattern of four circular gonads visible through the transparent bell. It has many fine tentacles around the margin of the bell. It is not considered dangerous. **4.** The By-the-wind Sailor, *Velella velella*, is a floating colonial organism related to hydroids (Phylum Cnidaria, Class Hydrozoa, p46). This species uses a sail to travel the oceans. The more harmful *Physalia utriculus*(next image) is closely related. **5.** Blue Bottles, *Physalia utriculus* (Phylum Cnidaria, Class Hydrozoa, p46) are superficially similar to the previous species but have a gas-filled bladder and will cause extremely painful stings from a single long blue tentacle. Extreme cases are rare but may cause breathing or cardiac difficulties—if these occur treat them as emergencies and seek immediate medical help. Hot water (NOT vinegar) is the recommended treatment. **6.** These little rocket-shaped gelatinous bells are siphonophores (Phylum Cnidaria, Class Hydrozoa, p46) which are more closely related to sea-floor dwelling hydroids than they are to other jellyfish.

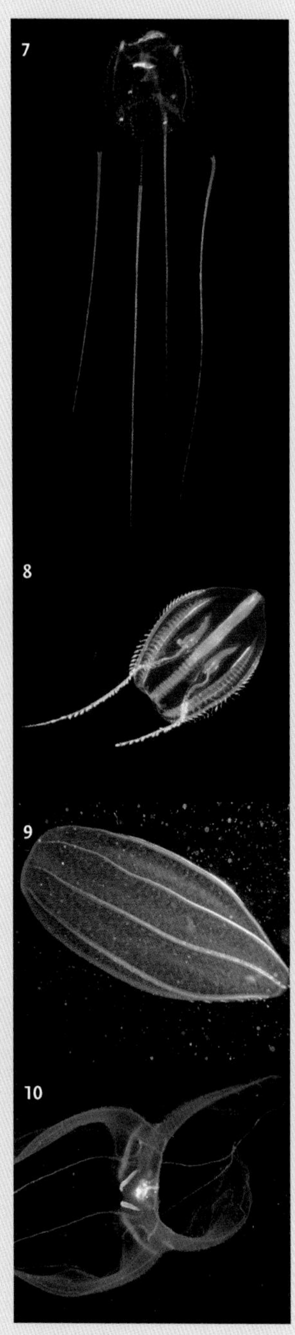

7. *Carybdea rastoni* is the only box jelly (Phylum Cnidaria, Class Cubozoa, p51) found in southern Australia. Other cubozoan species found in tropical waters can be fatal, but the sting of this species is merely painful. **8.** *Pleurobrachia* sp. is a comb jelly (Phylum Ctenophora) and is quite unrelated to cnidarian jellyfish. Comb jellies do not sting, most use sticky cells to catch their tiny prey. **9.** Members of the genus *Beroe* (Phylum Ctenophora, p53) are predators on other jellyfishes. They have a wide mouth and tooth-like lugs to tear off pieces of their gelatinous prey. The eight comb-like rows of cilia which characterise ctenophores are used for swimming and are clearly seen in this image. **10.** Other ctenophores (Phylum Ctenophora, p53) have strongly modified bodies with elaborate gelatinous lobes, yet the characteristic bands of cilia are still present. **11.** One of the most common groups of jelly-like planktonic creatures are the salps, oceanic sea squirts belonging to the Phylum Chordata (p109). This is a species of *Pegea*, which can exist as a solitary individual or in chains. **12.** Many salps (Phylum Chordata, p109) form long chains of repeated individuals, as in this *Pegea* species. Some can form wagon wheel circles. **13.** This salp (Phylum Chordata, p109) has the common name of a pyrosome. It is a species of *Pyrosoma*, a tube-shaped colony of numerous tiny individuals that is surprisingly rigid and robust. **14.** Some pyrosome salp colonies, such as *Pyrosoma spinosum* (Phylum Chordata, p109) are huge, reaching more than 10 m long. **15.** This firm, gelatinous kidney-shaped jelly is not a jelly fish but an egg mass produced by marine snails of the genus *Polinices* (Phylum Mollusca, Class Gastropoda, p84).

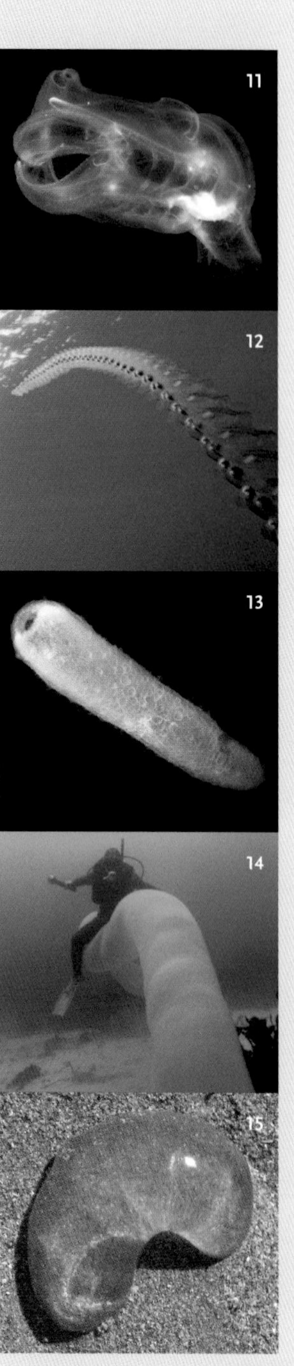

Quick Guide 5

Beach-washed remains and skeletons

The tide-mark on an ocean beach can be a rich treasure trove of mysterious skeletons, egg-cases and dead shells.

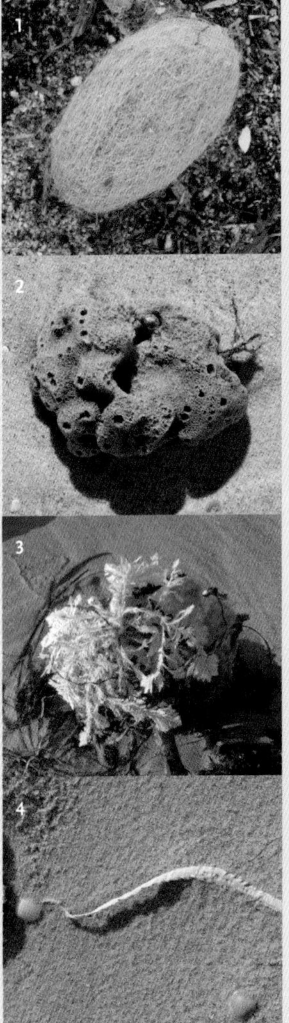

1. These balls are the fibrous remains of seagrass leaves (Phylum Anthophyta, p40) rolled into compact shapes by the action of waves. **2.** Sponges (Phylum Porifera, p42) are very common beach-washed objects. The bright colours typical of many living sponges are quickly lost and soon all that remains is the skeleton typical of a bath-sponge. **3.** These calcareous branches are the bleached skeleton of coralline red seaweed (Phylum Rhodophyta, p36). In life the seaweed is pink. **4.** These strange calcareous twigs are the skeleton of dead *Pseudogorgia godeffroyi*, corals related to sea anemones and gorgonian fan corals (Phylum Cnidaria, Class Anthozoa, p48). In life the colony lives attached to a shell or rock on the sandy sea floor. They wash ashore occasionally on ocean beaches. **5.** This strange sculpture is a lace coral (Phylum Bryozoa, p102). This *Celleporaria cristata*, is one of two species that envelop the stems of sea grass and are frequently beach-washed. **6.** Goose barnacles, *Lepas anserifera*, (Subphylum Crustacea, Class Cirripedia, p78) live attached to flotsam and jetsam and are therefore commonly found washed ashore. **7.** These tough parchment-like capsules are the eggs of a marine snail (Phylum Mollusca, Class Gastropoda, p84). **8.** *Sepia braggi* is a cuttlefish (Phylum Mollusca, Class Cephalopoda, p88) rarely found in inshore waters. However, the distinctive small narrow cuttlebone is commonly found washed ashore on ocean beaches in southern Australia. **9.** There are more than 10 species of

cuttlefish (Phylum Mollusca, Class Cephalopoda, p88) in southern Australia, and each has a different cuttlebone that floats ashore after the animal dies. This is from the largest species, *Sepia apama* (the living animal is shown on p89). **10.** The prize find of all beachcombers—the paper nautilus *Argonauta nodosa* —is actually the egg case of an open ocean swimming octopus (Phylum Mollusca, Class Cephalopoda, p88). Both animal and egg case are visible here. **11.** The Ram's Horn Shell is made by a deepwater squid, *Spirula spirula* (Phylum Mollusca, Class Cephalopoda). It is a common beach-washed object, especially along the east and west coasts of Australia. **12.** The 'test', or skeleton, of dead sea urchins (Phylum Echinodermata, Class Echinoidea, p98) often remain intact long enough to wash ashore. This is the test of a species of *Holopneustes*. **13.** Another sea urchin test (Phylum Echinodermata, Class Echinoidea, p98): *Heliocidaris erythrogramma*. **14.** The Ornate Cowfish, *Aracana ornata*, has a bony carapace and is therefore more often beach-washed intact than other fishes (Subphylum Vertebrata, Class Pisces, p115) **15.** The porcupine fish is another common beach-washed fish (Subphylum Vertebrata, Class Pisces, p115). Even after the body decomposes, the kidney-shaped swim bladder often remains. **16.** This distinctive helical twirl is the egg case of the Port Jackson Shark, *Heterodontus portusjacksoni* (Subphylum Vertebrata, Class Chondrichthyes, p112) a primitive bottom dwelling shark. **17.** These smooth, shiny egg cases are those of another elasmobranch (Subphylum Vertebrata, Class Chondrichthyes, p112), a skate (family Rajidae). **18.** Egg case of another elasmobranch (Subphylum Vertebrata, Class Chondrichthyes, p112), the elephant shark *Callorhynchus milii*. **19.** This thin collar-like structure is the egg mass of a marine snail of the family Naticidae (Phylum Mollusca, Class Gastropods, p84).

Quick Guide 6

Species commonly used as fishing bait

Fishers use all sorts of marine creatures as bait and some of the more commonly used species are included here. This section is intended to help anglers and others learn more about the plants and animals commonly used to lure fish.

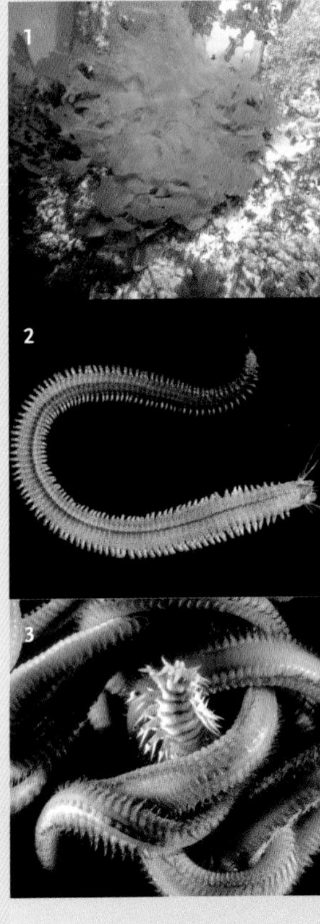

1. 'Sea lettuce', are green seaweeds (Phylum Chlorophyta, p34) in the genus *Ulva*. There are several species in southern Australia and they are favoured bait for herbivorous fish such as Luderick (*Girella tricuspidata*). **2.** Marine worms used for bait are all polychaetes (Phylum Annelida, p58). Sandworms of the family Nereididae (*Neanthes* spp. and *Perinereis* spp.) are found under stones in intertidal cobble. 'Pod Worms' *Australonereis ehlersi* are in the same family and are caught on sand flats in estuaries by digging or with bait pumps. *Neanthes* sp. is illustrated here. **3.** Beachworms (family Onuphidae) are also polychaetes (Phylum Annelida, p58) and occur on ocean beaches where they are caught by skilled collectors who lure them from the sand in the swash zone with fish heads or other carrion. There are several species, variously sold by bait shops as 'tube worms', 'sea worms' or 'bungum worms'. **4.** The most common bait species of pipi is *Donax deltoides*, which is widely distributed on ocean beaches around Australia, often found by wading the surf swash zone and feeling for the shells with bare feet. The shells have a distinctive purple interior and dead shells are frequently left behind by anglers. Pipis are also harvested for food, and along with many other molluscs, also occur in Aboriginal middens. Cockles (several species belonging to the genus *Katelysia*, and also *Anadara trapezia*) are also used as bait, as is the edible mussel common on wharf piles (*Mytilis galloprovincialis*). These are all bivalve molluscs (Phylum Mollusca, Class Bivalvia, p86). **5.** Octopuses (Phylum Mollusca, Class Cephalopoda, p88) are popular bait with line fishermen. In some regions, species such as *Octopus pallidus* (shown) are collected using PVC pipes with one end sealed with concrete. The octopuses use the tubes as shelter.

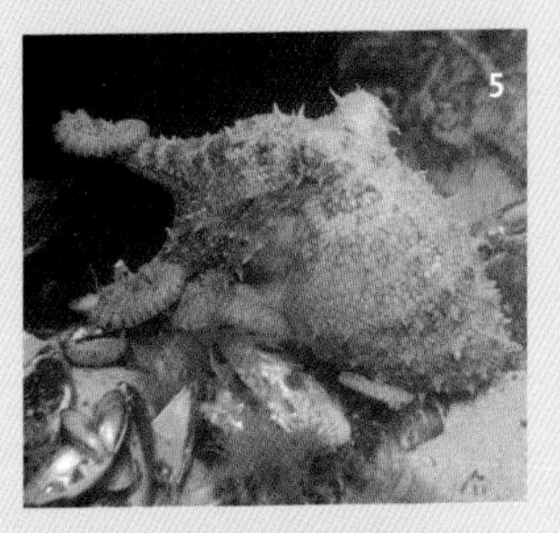

6. The Red Arrow Squid, *Nototodarus gouldi* (Phylum Mollusca, Class Cephalopoda, p88) is not only eaten by humans but is also used as bait. Calamari Squid (*sepiotenthis australis*) are often caught by jigging at night from piers, where they gather around lights. **7.** 'Ghost shrimp', 'nippers' and 'bass yabbies' are common names for crustaceans of the family Callianassidae (Phylum Crustacea, Order Decapoda, p76). Several species are used as bait and are collected with bait pumps on estuarine mudflats. **8.** A variety of crabs (Phylum Crustacea, Order Decapoda, p76) is used as bait but one of the most favoured is Red Bait Crab, *Plagusia chabrus*. This inhabitant of rocky ocean coasts is very fast moving and lives near low tide level—collecting these is often very dangerous and much more difficult than the subsequent catching of a fish! **9.** Soldier crabs, *Mictyris longicarpus*, are crustaceans (Phylum Crustacea, Order Decapoda) that are favoured bait for estuarine fish. They often occur in vast numbers on estuarine mudflats. **10.** Cunjevoi are sea squirts (Phylum Chordata, Class Ascidiacea, p108). The common species of ocean coasts in southern Australia is *Pyura stolonifera* which may almost be fist-sized and form extensive colonies near low water mark. Anglers cut away the tough, horny exterior to reveal the soft anatomy of the filter-feeding sea squirt, which is used as bait for reef-dwelling fish. **11.** Pilchards, *Sardinops neopilchardus*, are the most widely sold fish (Phylum Chordata, Class Pisces) used as bait. They occur in large schools in southern Australian waters, especially in summer months, when they form a major part of the diet of many fish and seabirds, including Little Penguins (*Eudyptyla minor*).

Quick Guide 7

Hazardous marine life—stings and venoms

Handling many marine animals may result in minor scratches or bites. Quick Guide 7 treats southern Australian marine species that cause more serious sickness, injury or fatality by stinging or venomous injection. Be aware that many other dangerous marine organisms are found in tropical waters. Some first aid advice is provided but this is not intended as a substitute for professional medical attention, which should always be sought whenever in doubt. See also Quick Guide 8 Hazardous marine life—poisons, bites and other dangers.

1. Blue Bottles, *Physalia utriculus* (Phylum Cnidaria, Class Hydrozoa, p46) cause extremely painful stings from their single long blue tentacle. Extreme cases are rare but may cause breathing or cardiac difficulties—if these occur, treat as an emergency and seek immediate medical help. Hot water (NOT vinegar) is the recommended treatment. **2.** The Lion's Mane Jellyfish, *Cyanea capillata* (Phylum Cnidaria, Class Scyphozoa, p50) can cause a painful sting and rash from a mass of trailing thin tentacles. For severe stings medical attention may be required. As with other cnidarian stings, vinegar is now not considered a useful treatment—it may cause attached tentacles to sting more vigorously. **3.** All cnidarians have stinging cells but most can not harm humans. Some, such as the Blue Blubber, *Catostylus mosaicus* (Phylum Cnidaria, Class Scyphozoa, p50) can cause a mild sting. **4.** Hydroids (Phylum Cnidaria, Class Hydrozoa, p46) also have stinging cells. Species found in southern Australia are generally not harmful but many are still able to sting bare skin. **5.** Cone shells are predatory gastropods (Phylum Mollusca, Class Gastropoda, p84) that capture fish and invertebrate prey by stabbing with a venomous dart on the end of an extensible proboscis. Some tropical species are extremely dangerous

and can be fatal. If necessary give resuscitation until the poison is broken down. *Conus anemone*, the common cone of southern Australia, is not known to cause fatalities but is best avoided. The proboscis can reach the back of the shell and the dart can penetrate gloves. **6.** Blue-ringed Octopuses, genus *Hapalochlaena* (Phylum Mollusca, Class Cephalopoda, p88) have been responsible for human fatalities. When hunting or resting, these octopuses are drab brown or grey. When disturbed they warn of their venomous nature with dramatic blue rings or lines, often against a dark spot or band. Their saliva contains tetrodotoxin, a powerful nerve toxin that blocks nerve transmission, causing paralysis and death by asphyxiation. Mouth-to-mouth resuscitation can keep a victim alive until the toxin wears off. All fatalities have occurred where people have handled these octopuses out of the water. **7.** Fire worms belong to the family Amphinomidae (Phylum Annelida, Class Polychaeta, p58) and have harpoon-like bristles, usually in prominent white bunches. The bristles are hollow and filled with toxins. Some species can cause painful stings and a lingering burning reaction (neoprene gloves offer no protection). Vinegar or anaesthetic creams can reduce the symptoms and sticky tape applied and removed over the sting may extract some bristles. **8.** The Glyceridae, sometimes known as blood worms, are a family of polychaete worms (Phylum Annelida, Class Polychaeta, p58) that burrow in sandy sediments. They have a muscular proboscis carrying four jaws that can be rapidly extended and can puncture skin with their bite. The jaws can inject a venom containing neurotoxins and may produce a local reaction lasting up to a few days. **9.** The spines of sea urchins (Phylum Echinodermata, Class Echinoidea, p98) can be very brittle and easily break off after puncturing the skin. Many species have venomous spines, especially tropical species. Southern species also should not be touched (especially the long-spined *Centrostephanus rodgersii*, which is venomous). Removing the spines is often a difficult and painful process and an infected wound frequently results. Hot water may reduce the immediate symptoms. **10.** The spines of many bony fishes (Subphylum Vertebrata, Class Pisces, p114) carry toxins. The toxins of scorpionfishes (including the tropical stonefishes) are susceptible to heat and break down at temperatures above 45°C. One treatment is to transfer the affected limb back and forth between heated and cold water. No temperate Australian fish is considered to have fatal toxins on its spines. **11.** As many anglers have discovered, flathead fishes have sharp spines that can cause painful wounds. Also be wary of Old Wives and wrasses. **12.** Stingrays are related to sharks in that both possess a cartilage skeleton (Subphylum Vertebrata, Class Chondrichthyes, p112). They get their name from one or more barbs on their tail. Stingray barbs have rows of small hooked teeth, the base of each containing a tiny pouch of toxin that ruptures when the barb is driven into an attacker. The barbs can cause painful ragged wounds. The effect of the toxin typically wears off after several hours.

Quick Guide 8

Hazardous marine life—poisons, bites and other dangers

Handling many marine animals may result in minor scratches or bites. Quick Guide 8 treats southern Australian marine animals that are potentially dangerous to eat, or that bite. Be aware that many other dangerous marine organisms are found in tropical waters. Some first aid advice is provided, but this is not intended as a substitute for professional medical attention, which should always be sought whenever in doubt.

1. Algal blooms or 'red tides' (Phylum Dinoflagellata and others, see p30) are blooms of minute single celled algae that under some conditions reach such abundances as to colour the sea red, yellow or brown. Swimming is not safe at these times, nor is eating seafood from the region, since the toxic cells accumulate in many marine organisms. Deaths have often occurred from eating seafood affected by red tides. Strictly follow health authority warnings. **2, 3.** Mussels and oysters are filter feeders which are able to concentrate some pollutants and toxins if they are present in the environment. Nowadays it is not advisable and often illegal, to collect shellfish for food close to large urban centres. Commercial mussel and oyster farms regularly monitor water quality. **4.** Sponges (Phylum Porifera, p42) contain minute spicules of silica and also a great variety of biochemicals. Handling sponges with bare skin can often cause discomfort and an irritating rash that may persist for days or weeks. **5.** Many pufferfishes (Subphylum Vertebrata, Class Pisces, p114) including the toadfishes, have flesh and organs that contain tetrodotoxin. This powerful toxin blocks nerve function and causes rapid paralysis and death. The pufferfish seafood dish in Japan, fugu, is prepared by specially trained chefs to ensure that all the poisonous flesh is removed, otherwise death

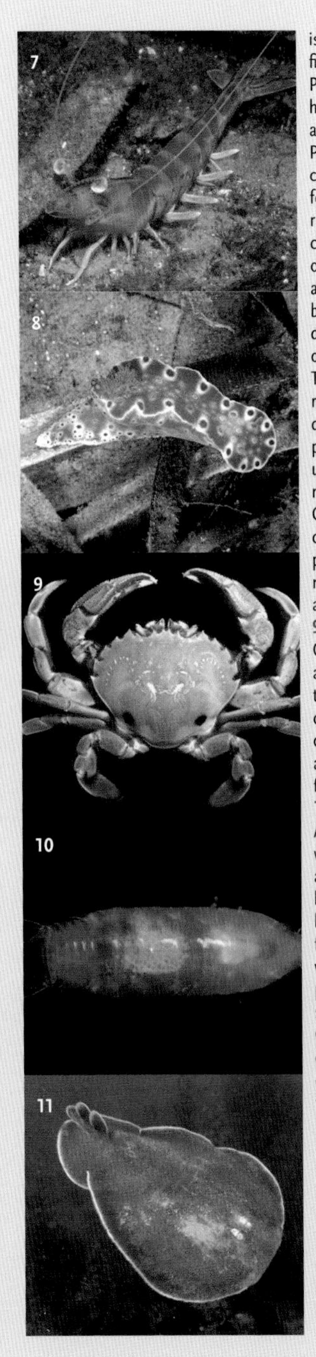

is a certainty. **6.** The porcupine fish (Subphylum Vertebrata, Class Pisces, p114) is closely related to the highly poisonous pufferfishes and are likely to be dangerous to eat. **7.** Prawns and other prized seafood can cause allergic reactions in an unfortunate few. Seafood allergies are reactions against proteins or other components and may be specific to one kind of seafood, or may apply to all products of the sea. People may become sensitised to seafood and develop an allergic reaction which once acquired is usually life-long. The symptoms may range from mild rashes to life-threatening breathing difficulties and drop in blood pressure. Serious cases require urgent medical treatment. **8.** Many nudibranchs (Phylum Mollusca, Class Gastropoda, p84) are brightly coloured, probably to warn potential predators that they are toxic and may be harmful to humans who are desperate enough to eat them. **9.** All crabs (Phylum Arthropoda, Order Decapoda, p76) have claws and most will use them to defend themselves. For example, the sand crab *Ovalipes australiensis* often lives concealed in sand on surf beaches and can deliver a painful nip to the foot of an unsuspecting swimmer. **10.** Sea lice are isopods (Phylum Arthropoda, Order Isopoda, p73) who scavenge the flesh of dead animals from the sea floor. They can be a nuisance to swimmers, biting bare skin as well as getting under face masks and wet suits of divers, where they can cause multiple painful bites. **11.** Electric rays are stingrays (Subphylum Vertebrata, Class Chondrichthyes, p112) that use electric shocks both to hunt and as defence. They take electrical charge from passing salt ions in seawater to gradually build up electricity in special capacitor organs in their flaps, like a trickle feed battery charger. They can voluntarily discharge shocks of up to 200 to 300 volts. **12.** The White Pointer or White Shark, *Carcharadon carcharias* (Subphylum Vertebrata, Class Chondrichthyes, p112) can reach six metres long. It normally feeds on seals, penguins and fish. This species has caused the most human shark attack fatalities. These tend to occur either near seal colonies or in the surf, where fleeing prey often hide from attackers. This species is protected in some regions as numbers are thought to have significantly declined from human fishing. **13.** The Wobbegong, *Orectolobus maculatus* (Subphylum Vertebrata, Class Chondrichthyes, p112) is so well camouflaged that a diver can easily rest on one by accident and receive a nasty bite. **14.** Many fish (Subphylum Vertebrata, Class Pisces, p114) have teeth that can cause a painful bite if given the opportunity. In particular, toadfishes, pufferfishes and leatherjackets (shown) have a wide fused beak instead of separate teeth, capable of a very strong bite.

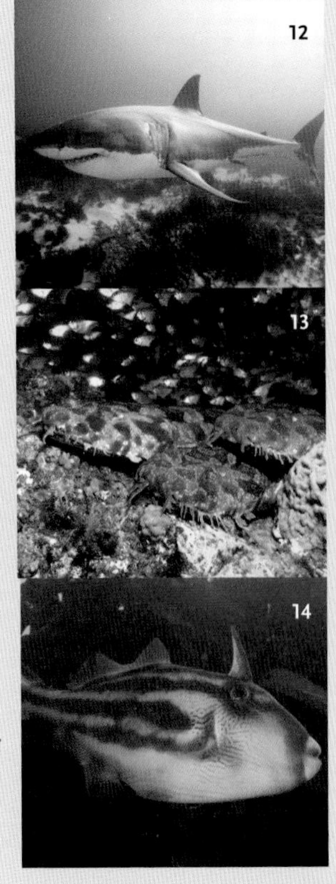

Quick Guide 9

Introduced species

Human activity, especially shipping, has caused the accidental spread of many marine species, both in Australia and throughout the world. As with weeds and animal pests on land, introduced species can result in significant changes to marine environments. There are hundreds of species of marine life now known to be introduced to Australian waters, plus many more whose identity we are still unsure of. Often they are quite similar to native species and accurate identifications may not be easy. Here we mention a few of the most obvious species in southern Australia. Many are also introduced to other locations around the world. It is important to clean gear and boats away from the sea and avoid further spreading these and other marine invaders.

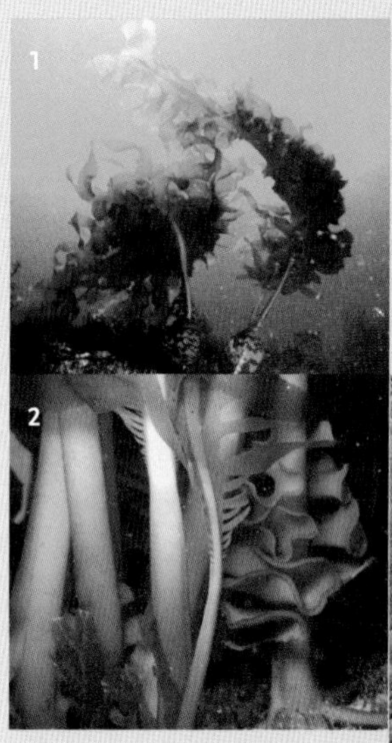

1, 2. *Undaria pinnatifida*, accidentally introduced to Tasmania and Victoria from Japan, is a brown alga (Phylum Phaeophyta, p38). The conspicuous 'ruffles' on the stalk distinguish this species from native kelps.

3. *Sabella spallanzanii* is a fan worm (Phylum Annelida, p58) accidentally introduced from the Mediterranean Sea and now widespread in ports and harbours of southern Australia. There are also large native fan worms but their filter crown is not in such a distinctive spiral.

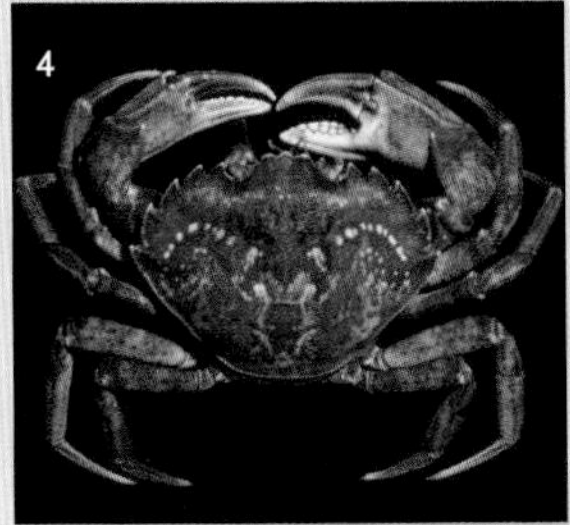

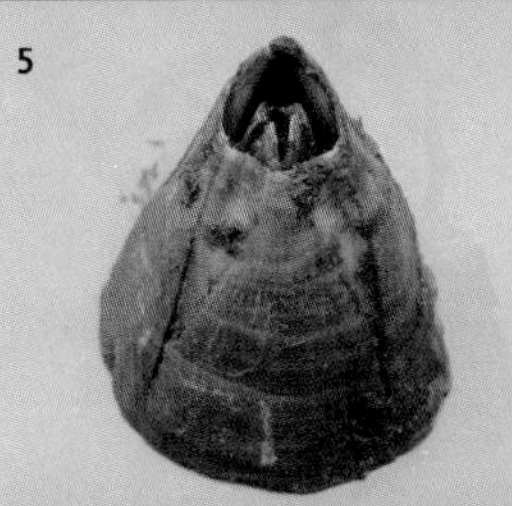

4. The European Shore crab, *Carcinus maenas* (Subphylum Crustacea, Order Decapoda, p76) was a very early arrival in Australian waters, where it arrived on ships' hulls by about 1900. Their greenish colour, shape and three notches on the carapace help to identify this species. **5.** *Megabalanus coccopoma* is a large barnacle (Subphylum Crustacea, Class Cirripedia, p78) that has been found in many Australian ports as a fouling organism on ships' hulls. This species appears to have reached Australia from its native Panama by ship. **6.** *Musculista senhousia* is a small mussel (Phylum Mollusca, Class Bivalvia, p86) endemic to asian seas but now spreading throughout southern Australia. **7.** The Pacific Oyster, *Crassostrea gigas* (Phylum Mollusca, Class Bivalvia, p86) is grown as an aquaculture species and eaten in restaurants throughout the world. Native to North Pacific shores, it has spread from Australian oyster farms and is now widely found in southern Australia. **8.** *Maoricolpus roseus* is a gastropod snail (Phylum Mollusca, Class Gastropoda) native to New Zealand. It was accidentally introduced to Australia among live oysters. **9.** *Asterias amurensis* is a sea star (Phylum Echinodermata, Class Asteroidea, p95) introduced to southern Australian ports from Japan. It has five arms, each with a purple, upturned tip, which distinguishes it from more evenly coloured native sea stars.

A Gallery of Marine Life

Lichens (Phylum Ascomycota)

Lichens are not marine, but some are maritime—that is, they live in the spray zone at and above high tide level. Lichens are actually a symbiotic association between a fungus and an alga, so each 'species' of lichen is more properly thought of as a partnership. Old textbooks classify lichens in their own group ('Mycophycota') but specialists now agree that they should be classified according to the fungal host. The fungal host of most lichens are from the Phylum Ascomycota. Lichens may take many forms, but the maritime species are all crusts that look like flaking paint, often in circles or arcs, formed as they grow outward from a central point. Thousands of species of lichen occur in Australia, but relatively few are typical of the maritime zone: *Parmelia* species are pale grey-green, *Xanthoria* species are usually yellowish, *Caloplaca* species are yellow to orange and *Verrucaria* species are black. However, this is only the roughest of guides; there are many more lichens and accurate identification requires specialist techniques (see Further Information).

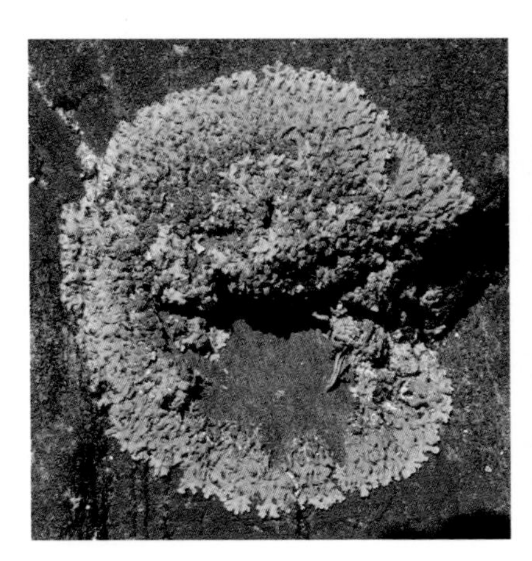

Dinoflagellates (Phylum Dinoflagellata) and other microalgae

Dinoflagellates are tiny single-celled organisms that swim using two whip-like tails called flagellae. They include both planktonic and benthic species, and their bodies are covered with ornamented plates made of cellulose. The life history of many dinoflagellates includes a dormant cyst phase which is resistant to unfavourable conditions. Some species live within other host animals, such as reef corals, molluscs or anemones, and use sunlight to make food for their host (photosynthesis). In this state, they lose their external plates and flagella. The largest dinoflagellate is about 1 mm long. There are more than 2,000 known living species, and another 2,000 species only known as fossil cysts.

Red tide microalgal bloom, Frecinet, Tasmania

Although dinoflagellates are too small to be seen individually, when conditions are right, they can breed rapidly and form large blooms called 'red tides' which appear as a yellow, red or brown scum floating on the surface of the sea. Some microalgal blooms are harmless, but others can be harmful or fatal to marine life and to humans who eat infected fish or molluscs.

Protoperidinium leonis (Congruentidiaceae)

Other kinds of single-celled organisms can also cause blooms, each with different environmental effects or problems for other marine life or humans. Collectively these organisms are referred to as 'harmful microalgae', which also includes diatoms (Phylum Bacillariophyta), raphidophytes (Phylum Chrysophyta), prymnesiophytes (Phylum Haptophyta) and blue-green algae (Phylum Cyanophyta—see below). Dinoflagellates cause paralytic, neurotoxic and diarrhetic shellfish poisoning and ciguatera fish poisoning, while diatoms cause amnesic shellfish poisoning. Humans, whales, dolphins, dugongs and seabirds can be poisoned. Humans are partly responsible for increasing the prevalence of harmful blooms, by raising nutrient levels (pollution), spreading cysts to new regions through shipping, and perhaps through climate change warming coastal waters.

Distinguishing different kinds of harmful microalgae is beyond the scope of this book, but it is important to be aware of the potentially dangerous effects of these microscopic organisms, and to heed any warnings from fisheries and health authorities (see Further Information).

Forams (Phylum Foraminifera)

Forams are very small and single-celled, and have shells with fine holes (pores) through which they extend filamentous 'legs' (podia). They include planktonic forms as well as benthic species that live on the sandy sea floor. Benthic forams use their legs to move slowly across the sea floor, or join together to form a net to catch passing prey. Forams grow by adding new chambers to their shell, and so they resemble a tiny spiral snail shell. Some forams are up to 2 cm in diameter, and are sometimes found by careful observers looking for small mollusc shells. There are some 40,000 known species of forams, many of which are fossils. Indeed, oceanographers have learned much about marine environments and past climates by studying the fossil record of these tiny but abundant organisms.

Unidentified Foraminifera

Blue-green algae (Phylum Cyanophyta)

Rivularia sp. (Rivulariaceae)

Symploca sp. (Phormidiaceae)

Blue-green algae are akin to bacteria—they have simple cells that lack structures such as a nucleus. Southern Australian blue-green algae form colonies on intertidal rocky shores and are often confused with true seaweeds. *Rivularia* species form dark green shiny cushions while *Symploca* species form low hairy irregular patches. Other blue-green algae communities form mounds known as stromatolites, like those at Shark Bay, Western Australia. They also occur as fossils up to 3.5 billion years old. Blue-green algae are thought to be among the oldest lineages of living things.

There are more than 2,000 known species of blue-green algae, including some from extreme habits such as hydrothermal springs and polar ice. Many are single-celled members of the plankton in marine and freshwater environments, where some may cause poisonous algal blooms (see discussion of Dinoflagellates, above). Other blue-green algae are commensals living within organisms such as plants, corals and lichen-forming fungi, where they generate energy through photosynthesis.

Stromatolites such as these at Hamelin Pools, Western Australia, are made by blue-green algae.

Green seaweeds (Phylum Chlorophyta)

The main photosynthetic pigment in green seaweeds is chlorophyll, causing their green colours. They are mostly found in the intertidal area and shallow marine habitats where light is abundant. If the water is very clear, they can occur as deep as 70 m. Green seaweeds have a wide range of forms: simple flat sheets, strings of beads, branching tubular structures, and forms such as the genus *Caulerpa* where a runner under the sand joins lines of upright fronds, almost like a seagrass.

IDENTIFICATION TIPS

Green seaweeds are most easily recognised by their green colour. Shape is the best way to identify species—such as whether it has flat thin fronds, long filaments, a large sphere or a branched form.

DIVERSITY

The Chlorophyta as a group contains more than 7,000 described species; many of these are microscopic, and many live in freshwater or terrestrial habitats, such as ponds and moist soil. More than 140 species of marine green seaweeds have been described from southern Australia.

Caulerpa brownii (Caulerpales)

ECOLOGY

Most green seaweeds live in the intertidal to shallow subtidal zones. They are typically attached to the substrate or other seaweeds and seagrasses. Different species are adapted for different environments, such as wave-exposed intertidal rock platforms, sheltered crevices or underwater reefs. Green seaweeds reproduce both sexually and asexually and may have complex life histories; different life stages may have quite different forms. The seaweeds provide habitat for a myriad of tiny marine animals. They are eaten by fishes, worms, echinoderms, crustaceans and molluscs.

Caulerpa flexilis (Caulerpales)

Codium lucasii (Codiales)

Cladophora prolifera (Cladophorales)

Dictyosphaeria sericea (Siphonocladales)

Ulva sp. (Ulvales)

Red seaweeds (Phylum Rhodophyta)

The deepest growing marine seaweeds are the red seaweeds. Like the other seaweeds, the red seaweeds contain green chlorophyll pigments, but also a range of other reddish pigments. The latter pigments allow the seaweed to collect light of a longer wavelength—the type of light that descends the deepest underwater. Although red seaweeds can also be found in the intertidal zone and on shallow subtidal reefs, their range extends deeper than the green and brown seaweeds, in some cases more than 200 m deep.

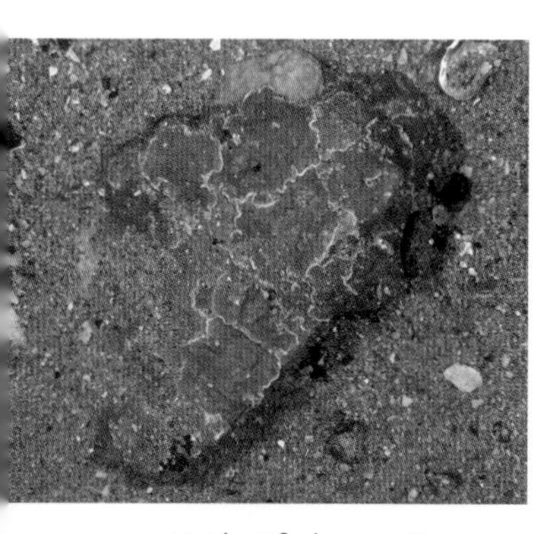

Unidentified encrusting coralline red alga

Red seaweeds are usually red in colour but (despite the name) they may also be yellow, brown, orange, purple or black. When some forms die or wash ashore they can bleach to become fluorescent pink or orange, eventually becoming white. There is great variety of form, from microscopic species, to filaments, to large forms—encrusting, flat and branching. Some red seaweeds are rigid and appear similar to corals; thus they are known as coralline algae. Their rigid structure is due to the calcium carbonate forming a limy skeleton in the cell walls. The coralline algae are an important component of intertidal and underwater reefs.

Champia viridis (Champiaceae)

Some species of red seaweed are important food sources for people. Nori (genus *Porphyra*) is particularly popular with Japanese people. Other species are commercially harvested for particular substances. The cell walls of red seaweeds contain two carbohydrates called agar and carrageenan, which are used in laboratory work, dairy products, cosmetics, textiles and medicines.

IDENTIFICATION TIPS

Shape is often useful for identification of species: fronds may be flattened, branched, filamentous or cylindrical. However, some species may be quite variable. For a definitive identification, it can be necessary to investigate finer details such as reproductive structures; this often requires a microscope.

DIVERSITY

Of all the seaweeds in southern Australia, the red seaweeds are the most diverse. There are more than 800 described species of red seaweed in southern Australia; worldwide there are more than 4,500 species. Almost all are marine. In Australia, there are also some species which have been introduced from other parts of the world, via shipping.

ECOLOGY

Red seaweeds live in a variety of habitats including rocky shores and underwater reefs. They are important hosts to many animals, such as small crustaceans. Sexual reproduction in red seaweeds is diverse; many different sexual cycles occur. They also increase through vegetative growth. Reproduction and other aspects of the biology of most southern Australian red seaweed species have never been studied in detail, and many discoveries undoubtedly await. This is also the case for other kinds of seaweeds.

Areschougia congesta
(Areschougiaceae)

Corallina officinalis
(Corallinaceae)

Dictyomenia harveyana
(Rhodomelaceae)

Brown seaweeds (Phylum Phaeophyta)

The largest seaweeds on temperate rocky shores are the brown seaweeds or Phaeophyta . These reach maximum size on cooler coasts, such as Victoria and Tasmania. Some of the larger species live at the low tide mark where there is often strong wave action. Subtidally, there are also large brown seaweeds such as the giant kelps, which may grow to 30 m long.

As with other seaweeds, the brown seaweeds have green chlorophyll pigments but they also possess pigments coloured yellow to red. These additional pigments capture more sunlight and so brown seaweeds can grow in relatively low light levels. Thus they can grow in deeper waters than green seaweeds, although not as deep as the red seaweeds.

Some species of brown seaweed are harvested for a substance called alginate, which is used in manufacturing food products and laboratory work. The whole seaweed plant may also be harvested for fertiliser. Certain species are eaten directly by humans, particularly in Japan.

Durvillaea potatorum
(Durvillaeaceae)

IDENTIFICATION TIPS

Brown seaweeds can usually be recognised by their colour, brown! The typical form is of a stalk (stipe) with attached flat blades, and a robust anchoring base (holdfast). There may also be gas-filled floats (vesicles) which help the seaweed to stay upright. Individual species vary in their shape; compare finely branched filaments versus flattened fronds, globular, branched fronds, the angle of the fronds to the stipe, and whether or not the stipe is straight or wavy. The life cycle of the brown seaweeds can involve different forms; thus, the appearance of some species might change depending on the age of the plant, and on the season. For example, a species may be small and encrusting in summer but by winter has long tubular fronds.

Hormosira banksii
(Hormosiraceae)

DIVERSITY

Worldwide, there are 1,500 described species of brown seaweed; about 240 of these are from southern Australia. The Phaeophyta are predominantly large marine plants, unlike the green seaweeds that include many microscopic and freshwater forms.

ECOLOGY

Brown seaweeds live on the intertidal rocky zone and subtidally to approximately 50 m deep. Some smaller species may grow on seagrasses or seaweeds. The fronds and the holdfast are homes to crustaceans, molluscs, sponges and worms. Some species are adapted for the degree of water flow, being specialised for rough or sheltered habitats. Growth rates may be high—up to 500 mm per day. Reproduction, as with other marine macroalgae, occurs through diverse strategies. Brown seaweeds employ various modes of sexual reproduction (release of eggs and sperm), but asexual reproduction also occurs through release of asexual spores.

Sargassum sp. (Sargassaceae)

Ecklonia radiata (Alariaceae)

Zonaria sp. (Dictyotaceae)

Seagrasses and mangroves (Phylum Anthophyta)

Avicennia marina (Verbenaceae)

Amphibolis antarctica (Cymodoceaceae)

Halophila ovalis (Hydrocharitaceae)

The angiosperms, or flowering plants, are the most common land plants. Mangroves and seagrasses are among the few angiosperms that occur in marine environments. Mangroves are trees of intertidal estuarine mud flats, while seagrasses form grass-like meadows in shallow bays and estuaries. A diverse flora of flowering plants (as well as some mosses, ferns and allied plants) occur on sand dunes and salt marshes—these plants may extend down to high water mark but these environments are beyond the scope of this book.

IDENTIFICATION TIPS

Mangroves are shrub-like trees living on mudflats surrounded by numerous air-breathing stalks (pneumatophores). Seagrasses are green and might be confused with green seaweeds (algae). Like flowering plants on land, seagrasses also reproduce by seeds, but these are not easily seen. However, seagrasses have more complex structures: leaves, stems and rhizomes (roots). Cutting through a seagrass stem will reveal sections of tubular vessels which carry fluids and nutrients. Seaweeds lack these internal structures.

DIVERSITY

A single species of mangrove (*Avicennia marina*) and seven seagrass species are known from Victorian waters. Many more species of both groups occur in tropical Australia, and many more kinds of flowering plant can be found in saltmarshes and other maritime environments.

Mangrove trees are only found in sheltered intertidal environments, typically in muddy estuaries. Their biology adapts them to life in an environment unsuitable to most flowering plants, where the water is salty and the sediment lacks oxygen. Mangrove roots have vertical extensions into the air—pneumatophores—that compensate for the lack of oxygen in the sediment. Mangrove seeds germinate while still attached to the plant; when they detach they float, allowing dispersal to a new location.

Seagrasses occur on sandy and muddy sediments, usually subtidally, although species of eelgrass (*Zostera* and *Heterozostera* species) are more shallow, and *Zostera* beds are regularly exposed at low tides. Sea Nymph (*Amphibolis antarctica*) is found in clean sand on the open coast, and often has epiphytic red seaweeds growing on the stem and leaves. Paddleweeds (*Halophila* species) occur in slightly deeper water, to more than 10 m deep. Seagrass flowers are small and difficult to see; pollen is released from the male flower and drifts through the water to fertilise female flowers. So far as is known, there is no marine animal that assists dispersal of the pollen and fertilisation. Asexual reproduction also occurs, typically resulting in extensive submarine seagrass meadows.

Heterozostera sp. (Zosteraceae)

Posidonia australis (Posidoniaceae)

Sponges (Phylum Porifera)

Sponges are the simplest of all multicellular animals. They are comprised of tubes and chambers lined by specialised cells with hair-like flagellae. The beating flagellae drive water through the chambers to capture food particles. Most sponges are supported by crystalline rods known as spicules. The mineral composition of the spicules defines the three kinds of sponges. The Demospongiae have silica spicules that are simple rods or 4-rayed; these sponges are the most common and diverse shallow water sponges. Calcareous sponges have spicules of calcium carbonate. Glass sponges (Hexactinellida) have 6-rayed silica spicules and are primarily deep-water organisms.

Dendrilla rosea (Aplysillidae)

Sponges on pier piling

IDENTIFICATION TIPS

Sponges are best thought of as loose aggregations of individual tiny creatures, rather than the highly structured colonies which make up corals, hydroids, bryozoans and some sea squirts (colonial ascidians). Some colonial ascidians resemble sponges but can be differentiated by their paired feeding pores and their internal specialisations.

DIVERSITY

Worldwide there are over 6,000 described species of sponges. Over 1,000 species have been described from Australia. Only a few species of sponges are sufficiently distinct that they can be easily recognised and most require specialised techniques for study (such as dissolving the flesh in acid and looking at the spicules under a microscope).

ECOLOGY

In shallow waters well lit by sunlight, seaweeds usually out-compete sponges, but in deeper water, in caves and under dark piers sponges can be the dominant encrusting organism. They thrive where high water movement provides a reliable food source, as on vertical walls or in channels. Reproduction can be sexual, or asexual—by forming of a new sponge from a fragment. Most species are hermaphrodites, producing eggs and sperm, but at different times. Fertilisation is usually external, with the developing embryo having a planktonic existence, however internal fertilisation of the egg with subsequent release of an embryo is also known to occur.

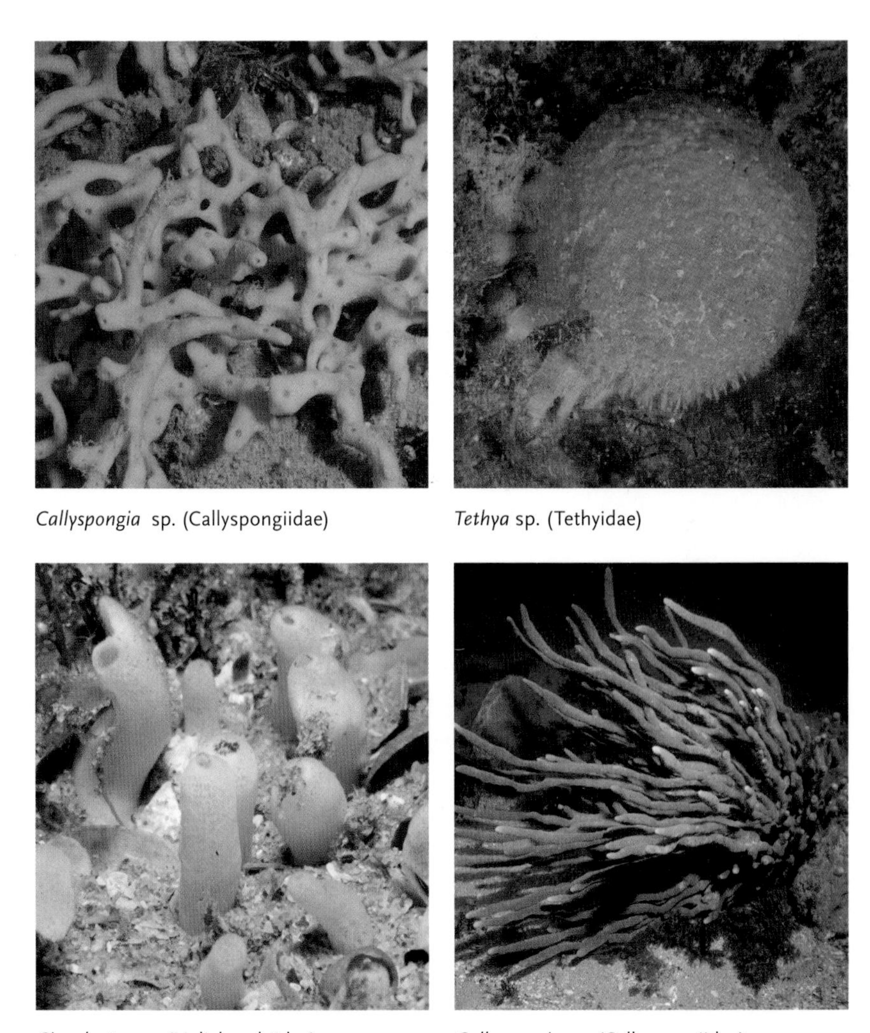

Callyspongia sp. (Callyspongiidae)

Tethya sp. (Tethyidae)

Ciocalypta sp. (Halichondriidae)

Callyspongia sp. (Callyspongiidae)

Gorgonian fan coral (Melithaeidae)

Corals and jellyfishes (Phylum Cnidaria)

The cnidarians include many of the most familiar sea creatures: jellyfishes, anemones, corals and their relatives. These related animals all share stinging cells called 'nematocysts'. These cells contain coiled barbs and toxins that are released when prey brushes against the cell. Also common to these animals are bodies which are circular or with a star-like shape ('radial symmetry'). This radial structure is easily seen in a jellyfish, coral polyp or sea anemone. Cnidarians have a simple basic body plan: mouth, stomach and tentacles. The life cycle of many cnidarians includes both a floating jellyfish stage (known as a medusa), and attached anemone-like form known as a polyp. In some kinds of cnidarians (such as jellyfish) the medusa is the free-floating form familiar to us, whereas in others, for example the hydroids, there is an attached polyp. However, there are many variations on life-history, and in corals, anemones and relatives there is no medusa phase at all. Many cnidarians are also able to reproduce by 'budding' new individuals, thus forming a large colony, as in reef-building corals and in feather-like hydroid colonies.

Parazoanthus sp. (Parazoanthidae)

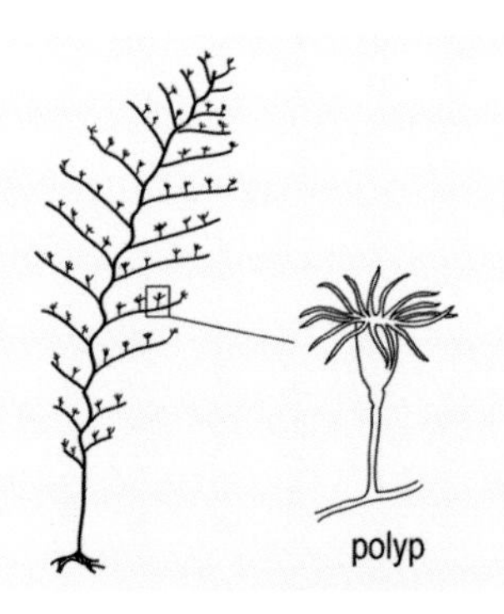

Generalised hydrozoan

Gymnangium superbe (Aglaopheniidae)

Ralpharia magnifica (Tubulariidae)

Hydrozoans

(Phylum Cnidaria, Class Hydrozoa)

The most commonly seen hydrozoans in temperate seas are hydroids (Subclass Leptolida) which form branching colonies or stalked polyps attached to rocks, kelp and the like. Most hydroids comprise many minute polyps making up a colony that is often delicate and fan- or feather-like. A few species have larger polyps on long erect stems, like a jellyfish on a stick. Like all cnidarians, hydroids have stinging cells (nematocysts), as is well known by divers who have brushed them with bare skin! Hydroid colonies grow larger by asexual budding of new polyps, but they can also reproduce sexually with release of mobile medusae (like tiny jellyfish) into the water.

Other hydrozoans form floating jellyfish-like free-swimming colonies with polyps modified for different functions. These floating hydrozoan colonies typically have a gas float or muscular pumping structure at one end and a branch of dangling tentacles trailing behind, armed with powerful stinging cells. Commonly seen examples are the Blue Bottles (*Physalia utriculus,* belonging to the Order Siphonophorae, also known as siphonophores) and the By-the-wind Sailor (*Velella velella,* belonging to the Order Anthomedusae).

Experts differ in their views of the classification of the Hydrozoa, as can be seen by consulting the sources listed in Further Information.

IDENTIFICATION TIPS

Hydroid colonies are easily confused with other kinds of colonial organisms, especially lace corals (bryozoans, p102). Hydroid colonies are made of chitin and are often brownish and flexible, while bryozoan colonies are usually calcified and brittle. Crush a branch with metal forceps: if it makes a 'crunching' sound it is probably a bryozoan; otherwise it is probably a hydroid. The most commonly-seen hydroids are often flattened two-dimensional colonies, whereas bryozoans are mostly bushy and three-dimensional. Several species of hydroid form fan-like colonies which resemble gorgonian corals (fan corals, p44). However gorgonians are mostly red or orange, not the typical brown colour of a hydroid colony.

Solanderia fusca (Solanderiidae)

DIVERSITY

Several hundred hydroid species occur in southern Australian waters, and many more small or deep water species undoubtedly await description. A few common and distinctive species can be easily recognised, but accurate identification of most hydroid species is based on microscopic structures of the polyps and nematocysts.

Macrorhynchia whiteleggei (Aglaopheniidae)

ECOLOGY

Hydroids are static predators which use their stinging cells to capture and kill planktonic organisms. Since a hydroid colony is fixed, it must rely on wave action or currents to bring the tiny planktonic prey past the polyps. Therefore they are not often found in still waters. Humans can be stung by hydroid nematocysts. Some species in tropical waters cause very painful stings, but, at most, some southern species can cause a mild tingling sting if grasped with bare hands.

Velella velella (Porpitidae)

Some of the free-swimming forms like the siphonophores cause a painful sting that occasionally requires medical attention. As they can wash ashore on ocean beaches they can be a hazard to swimmers, surfers and snorkellers.

Physalia utriculus (Physaliidae)

Sarcoptilus sp. (Pteroeididae)

Anthothoe albocincta (Sagartiidae)

Unidentified gorgonian (Melithaeidae)

Sea anemones and corals

(Phylum Cnidaria, Class Anthozoa)

Sea anemones, corals, and sea pens—anthozoans—belong to the Phylum Cnidaria, and typically live attached to the sea floor or some other substrate. Anthozoan polyps are usually larger and more complex than a hydroid polyp and have internal connections (mesenteries) arranged like the spokes of a wheel.

The most familiar anthozoans in temperate seas are sea anemones. Corals are also anthozoans, including the gorgonian (fan) corals and soft corals often seen by divers. Stony corals are more common in tropical waters where they form coral reefs, but a number of less conspicuous species also occur in southern Australia. The least familiar anthozoans are sea pens, which are fleshy leaf-like colonies of highly modified polyps arranged in two rows around a supporting stem.

IDENTIFICATION TIPS

Corals and anemones are distinctive and are unlikely to be confused with any other marine organisms. Gorgonians may resemble some fan-like hydroid colonies (see p45). Soft corals and sea pens are more recognisable when seen alive and undisturbed, but when washed ashore may look like mysterious rubbery objects.

DIVERSITY

Only about 200 species of anthozoans are known from southern Australia. However, Australian anthozoans have not been well studied and many species await description.

ECOLOGY

Most anemones and corals need to attach to rock or another hard surface, however a family of small anemones (the Edwardsiidae, or burrowing anemones) live in soft sediments. Sea pens also live on sandy bottoms, where they are anchored by the buried bottom part of the stem. Gorgonians and soft corals are more abundant in deeper water, or in darker caves and crevices in shallow water. Virtually all anthozoans prey on small planktonic organisms, so all require good water movement. Anthozoans do not have a mobile free-swimming (medusoid) stage in their life cycle; instead dispersal is achieved through release of tiny young (called planula larvae).

Unidentified soft coral (Nephtheidae)

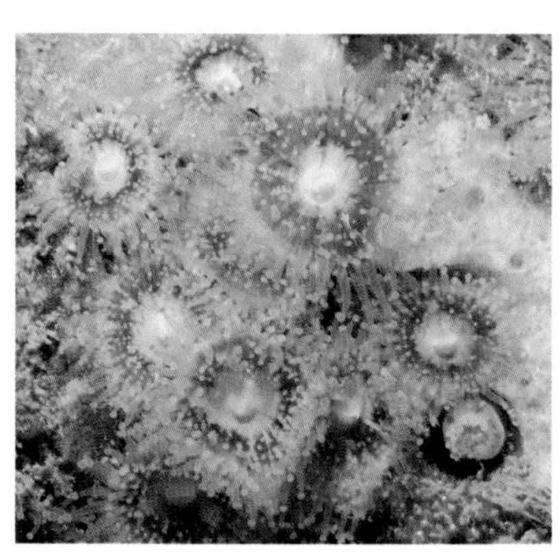

Corynactis australis (Aurelianidae)

Phlyctenactis tuberculosa (Actiniidae)

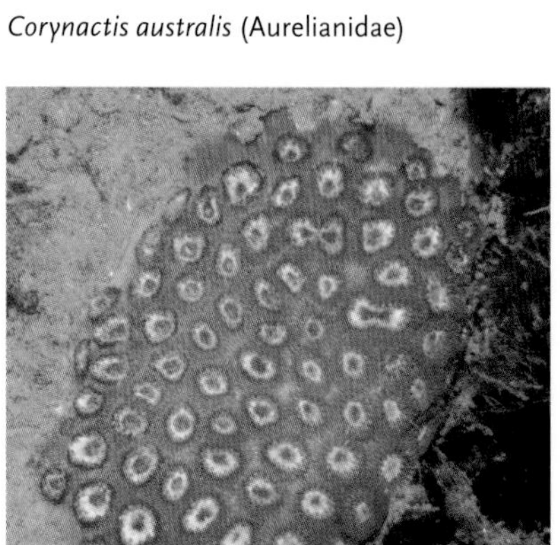

Plesiastrea versipora (Faviidae)

Primnoella sp. (Primnoidae)

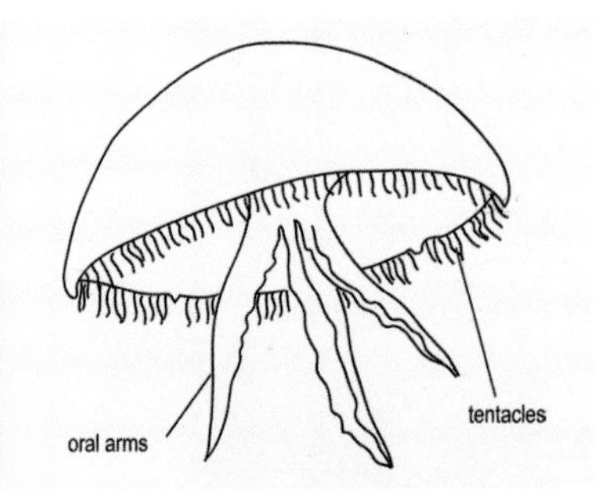

Generalised scyphozoan

Cyanea annaskala (Cyaneidae)

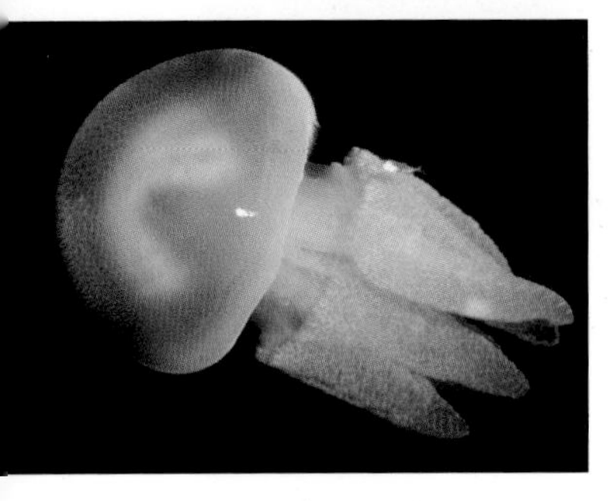

Catostylus mosaicus (Catostylidae)

Jellyfish

(Phylum Cnidaria, Class Scyphozoa)

Jellyfish are floating gelatinous sea creatures related to corals and anemones. Most of the large jellyfish in southern Australian waters are members of the Scyphozoa. They are often umbrella- or bowl-shaped free-swimming creatures, typically armed with tentacles and stinging cells (nematocysts). Jellyfish live in the water column and typically feed by capturing small animals on their trailing tentacles. They are mostly encountered as jelly blobs washed up on beaches or as mild irritations from stings when swimming.

IDENTIFICATION TIPS

Scyphozoan jellyfishes have a rounded symmetrical bell typically with a fringe of tentacles. The gonads and lobes of the stomach can be seen through the bell of many species, typically in a four-rayed arrangement like a cross or a lucky clover leaf. Most species have oral arms (typically four) hanging from the inside centre of the bell. Some species have additional filaments or a long tail extending from the oral arms. One scyphozoan group, the Stauromedusae ('stalked jellyfish'), is not free-swimming but spend their life attached to weed or rocks, looking like a skinny anemone.

DIVERSITY

There are around 250 species of scyphozoan jellyfishes in the world, all of which are marine. Because of their lifestyle of travelling the world's ocean currents, many species occur all around the world. Around ten species are found along the south coast of Australia.

ECOLOGY

Scyphozoan jellyfishes capture their food using nematocycsts. These are typically on the tentacles

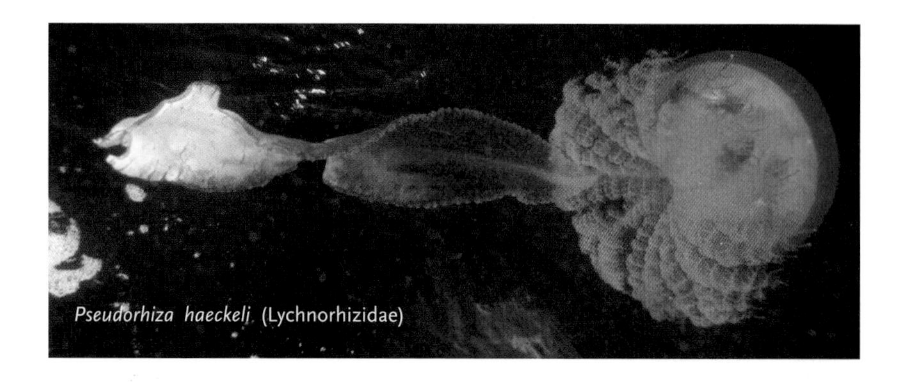
Pseudorhiza haeckeli (Lychnorhizidae)

but can also be on the bell, the oral arms or a tail . Most scyphozoan jellyfishes breed by having a tiny bottom-dwelling stage, known as the scyphostoma. This is a small slender polyp that produces tiny jellyfish in stacks like a pile of small dinner plates. Each one buds off the end and swims away to form the adult jellyfish.

Box jellies

(Phylum Cnidaria, Class Cubozoa)

Tropical Australian waters are famous for their deadly box jellies. The southern Australian member of this group, *Carybdea rastoni,* is less well known as its sting is not deadly. Box jellies are strong swimmers using regular beats of their muscular cube-shaped bell. Retractable tentacles are attached at each corner of the cube. There are around forty species of box jellies worldwide. The local species gathers in large groups during the day, dispersing to feed at night by swimming regular diagonal passes from the sea floor to the surface.

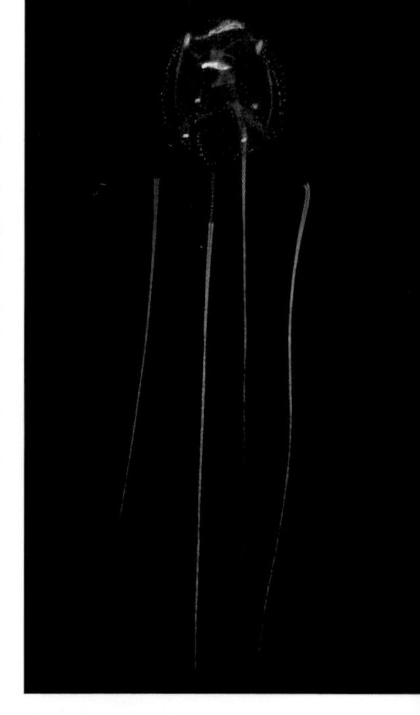
Carybdea rastoni
(Carybdeidae)

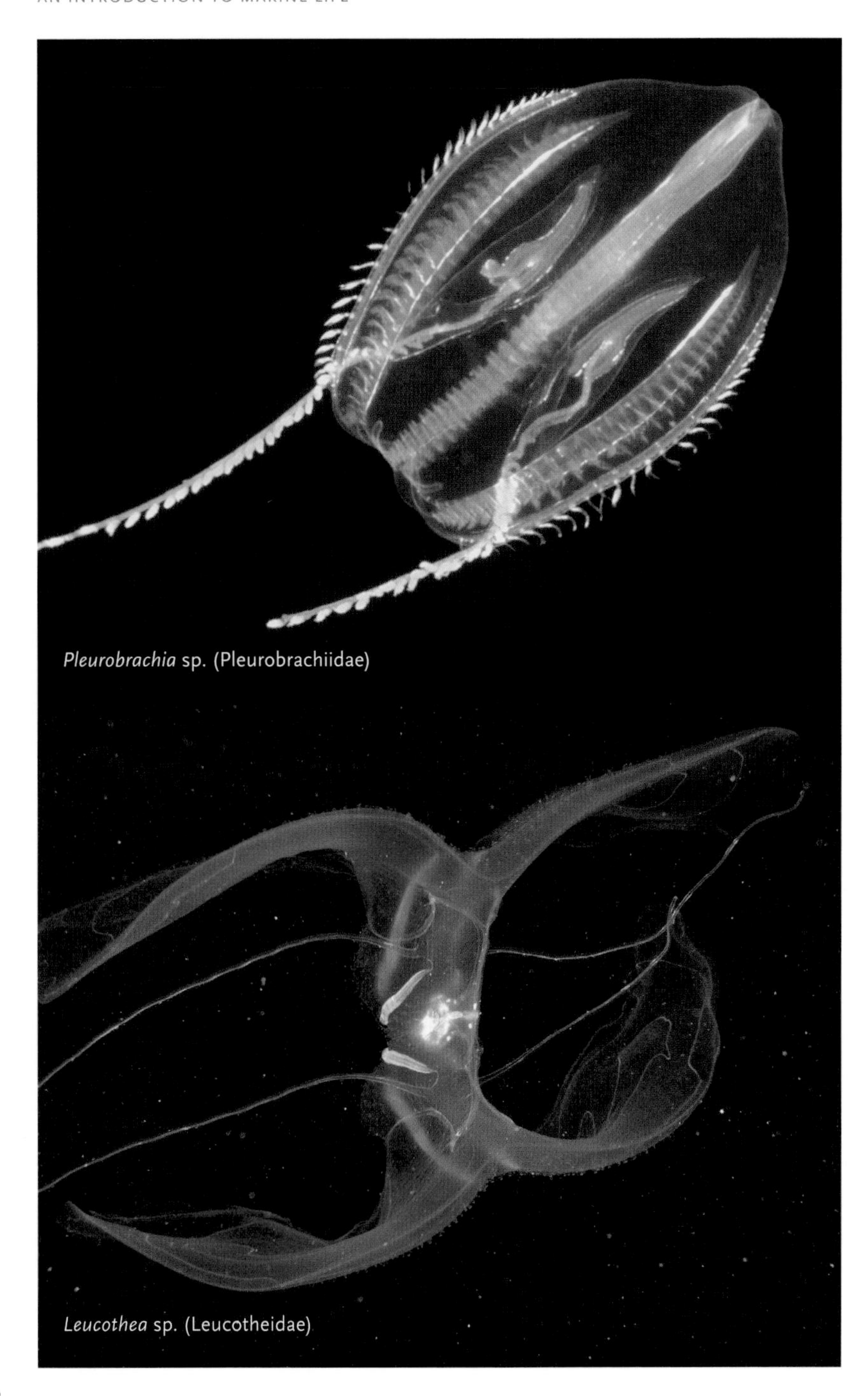

Pleurobrachia sp. (Pleurobrachiidae)

Leucothea sp. (Leucotheidae)

Comb jellies (Phylum Ctenophora)

Comb jellies are a special group of transparent jelly-like animals that are placed in their own phylum. Their name comes from the rows of fine beating hairs that they use for locomotion. These hairs refract the light to make beautiful pulsing rainbows, making these delicate animals some of the most beautiful of all the sea creatures. Ctenophores are, however, not easily found as they are typically small, transparent and difficult to see against bright sunlight at the surface. The best time to see them underwater is at night with the aid of dive lights. They can occur in large numbers, and sometimes swarm in quiet Port Phillip Bay waters in summer.

IDENTIFICATION TIPS

Comb jellies have jelly-like bodies and their 'combs': the eight rows of small beating hairs by which they travel. Most species have an egg-shaped body with long, very active retractable tentacles. There are also other forms including a long blade-like form known as a Venus' Girdle, a winged form known as a lobate comb jelly, and even creeping forms that live on the seafloor.

DIVERSITY

There are around 100 species of comb jellies worldwide. The number of species in southern Australia is not known. The different species are difficult to tell apart. There are also many deep-sea species.

ECOLOGY

Most comb jellies feed using tentacles that are armed with adhesive cells that grip their prey like glue. Some can catch moderate-sized crustaceans and even small fish. Other comb jellies prey on jellyfish. *Beroe* species (see Quick Guide 4, image 9) have a wide mouth armed with small tooth-like lugs. It sucks jellyfish into its stomach and saws off the bits that are sticking out—they are like the shark of the comb jellies. Those ctenophores which live on the bottom have flattened bodies and reel out their sticky tentacles to trap passing prey. Ctenophores are typically hermaphrodites and release eggs and sperm into the water where fertilisation occurs and the embryo develops into a planktonic larva. Benthic species, however, have internal fertilisation and subsequent release of the developing larva.

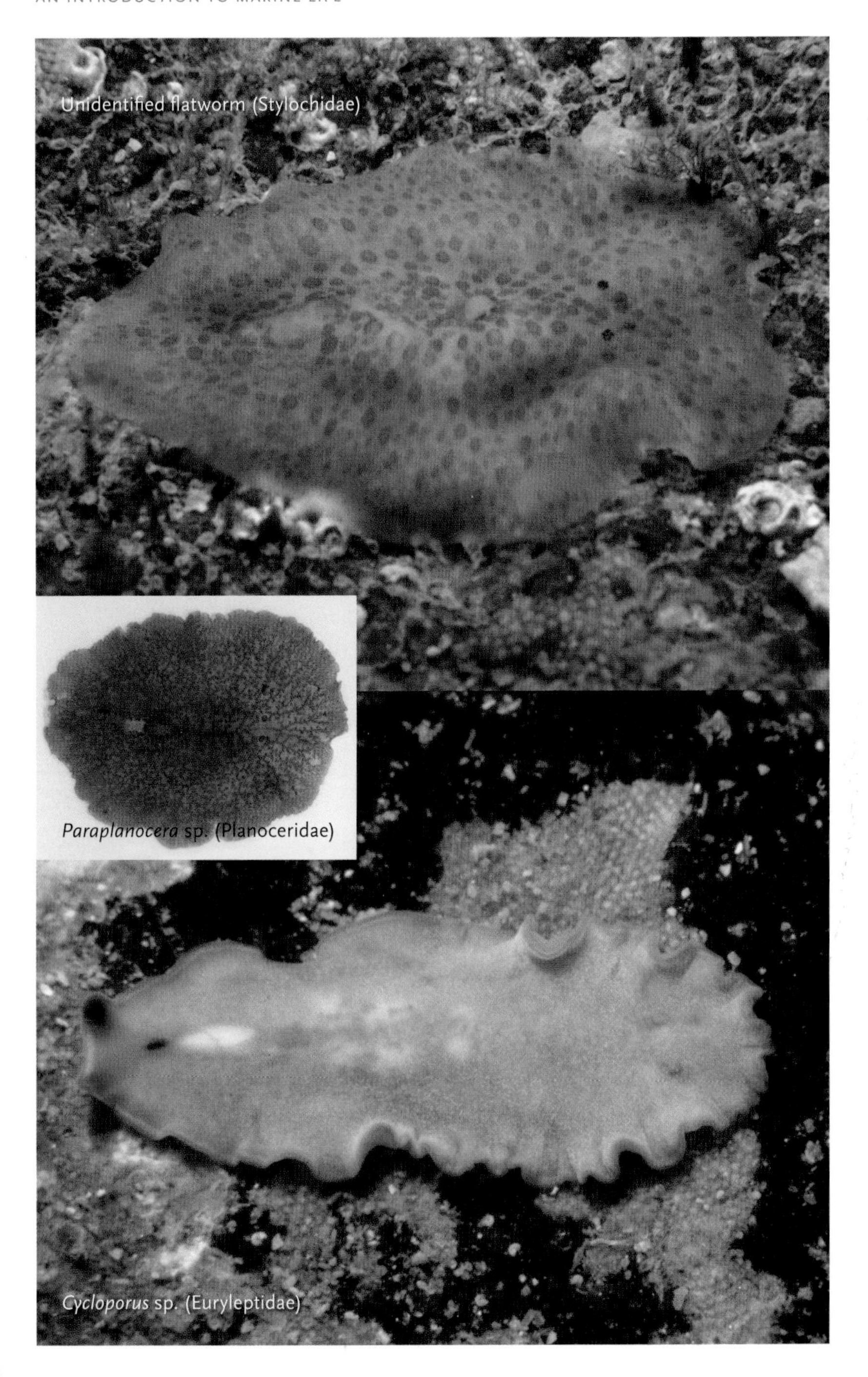

Unidentified flatworm (Stylochidae)

Paraplanocera sp. (Planoceridae)

Cycloporus sp. (Euryleptidae)

Flatworms
(Phylum Platyhelminthes)

As their name suggests, flatworms are thin, flat, and wide; most are short and oval but some forms are long. Many kinds of flatworms are parasites or tiny creatures living between sand grains and are thus rarely seen. They have unsegmented, simple bodies with no obvious appendages. There is no gut cavity and instead intestinal branches radiate from a central mouth and pharynx (a muscular organ used to engulf prey). The most commonly observed flatworms are the polyclads (Polycladida), which may be several centimetres long and either brown or brightly pigmented. Triclads (Tricladida) are also abundant and diverse in marine (and other) environments, but are mostly minute and not often seen. Polyclads have guts with many branches, while triclads have a gut with three branches.

IDENTIFICATION TIPS

Flatworms can usually be distinguished from other worms by their thin, flat body. The only other flat bodied worms are ribbon worms (nemerteans), but they are much longer than flatworms. Live flatworms can swim and crawl quickly, whereas live nemerteans usually cannot. Brightly coloured species sometimes look very like nudibranchs (molluscan sea slugs). However, turn over a nudibranch (which is easy to do—due to their slow and methodical approach to life) to see underneath a separate muscular foot. The underneath of a flatworm is simple and flat (by comparison they are hyperactive and resist interference).

DIVERSITY

At least 10,000 species of Platyhelminthes are known worldwide, but only twenty-five or so are recorded from southern Australia. Clearly there are many more to be discovered and described.

ECOLOGY

Polyclad flatworms are active predators which crawl over their prey, stick out their pharynx, and digest the prey externally. In tropical waters, some polyclads gain protection by mimicking the colour patterns of toxic nudibranchs, but it is not known if similar mimicry occurs among the southern fauna. Polyclads are hermaphrodites and conduct sex through a variety of arcane practices, often involving 'penis fencing' where the worms stab each other to introduce sperm. (see Further Information).

Ribbon worms (Phylum Nemertea)

Ribbon worms reach lengths of many metres and resemble long, flat ribbons. They lack a gut cavity but have extendable mouthparts (called a proboscis), digestive tract, blood vessels and nervous system. The head end often has long slits resembling a smiling mouth, these are actually sensory structures. Ribbon worms are often flesh-coloured, some species have distinctive colour patterns, often longitudinal stripes.

Notospermus geniculatus
(Lineidae)

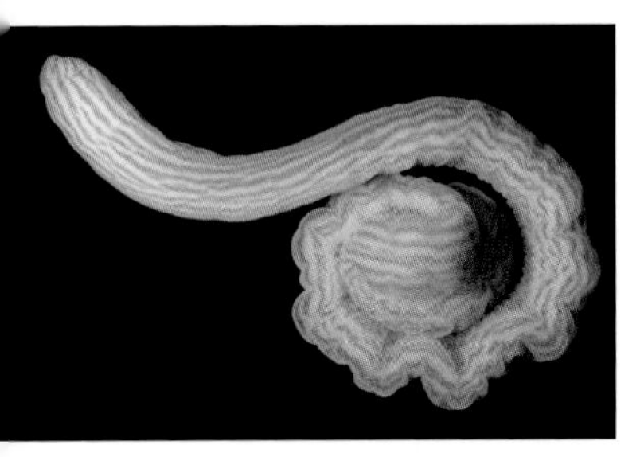

Baseodiscus delineatus
(Baseodiscidae)

IDENTIFICATION TIPS

Ribbon worms are unsegmented, like flatworms, but tend to be rounder and typically much longer. Unlike flatworms, which have a single orifice which is both anus and mouth, nemerteans have a separate mouth and anus. Ribbon worms are very fragile and although widespread, they are a challenge to collect intact and even more difficult to identify. Colour photographs of live specimens are the best way to learn about the local fauna.

DIVERSITY

Over 1,000 species of nemerteans are known globally, but these animals are difficult to study and fewer than 40 species are presently recorded from southern Australia. Many new species undoubtedly await discovery and description.

ECOLOGY

Most nemerteans are predators living under stones, in algae, and among encrusting invertebrates. However some are commensal or parasitic on other invertebrates such as ascidians and crustaceans. Many nemerteans can regenerate from fragments, and some reproduce asexually in this way, each fragment growing into a new worm. Sexes are separate, although some are hermaphrodites. Eggs and sperm can be released into the water, where fertilisation occurs and a planktonic larva develops. Others copulate and fertilise eggs and develop larvae internally.

Peanut worms (Phylum Sipuncula)

Peanut worms or sipunculans are unsegmented with a somewhat bottle-shaped body, bulbous at one end and with a slender extendible section called an introvert. The mouth is at the end of the introvert and leads to a strongly coiled intestine and anus located on the side of the body. There are muscle bands extending through and around the body, so that peanut worms can retract and expand readily. The body surface is often speckled with either papillae or pores and is often distinctively 'leathery' in feel.

Generalised peanut worm

IDENTIFICATION TIPS

Sipunculans are easy enough to distinguish from other kinds of unsegmented worms: Spoon worms (specialised members of the Phylum Annelida) are green, soft and sluglike with long, t-shaped mouthparts. Flatworms are very flat. Ribbon worms (Phylum Nemertea) are usually long, ribbon-like and fragile. Penis worms (Phylum Priapula p61) have a body that is divided into two distinct regions, and appendages at the tail end that look like bunches of grapes. The leathery, muscular sipunculans should be easy to recognise.

Phascolosoma sp.
Phascolosomatidae)

DIVERSITY

The sipunculans are not very diverse, only about 150 species have been described worldwide; of these, about forty-six are known to occur in Australia.

ECOLOGY

Sipunculans are all burrowers; mostly in sand and mud, but some (e.g., *Themiste fusca*) bore holes in limestone rock. Several species of *Phascolosoma* do likewise but are also found under rocks and in colonies of the intertidal encrusting polychaete worm *Galeolaria caespitosa*. Species of *Golfingia* occur in sediment, especially among seagrass roots. The diet of most sipunculans probably consists of small particles of detritus. Sexes are usually separate with external fertilisation.

Segmented worms (Phylum Annelida)

Any worm made up of numerous segments is probably an annelid. In the sea, the most commonly seen annelids are polychaetes (meaning literally 'many chaetae', referring to the bristle-like chaetae projecting in bundles from each segment). Polychaetes are known to anglers by names such as sandworms, lugworms and bloodworms. Other kinds of annelids are oligochaetes and leeches. Most marine oligochaetes are tiny thread-like worms which can be very numerous but are only noticed by ecologists sampling with fine sieves. There are a few marine leeches, mostly found on sharks and rays.

Lepidonotus melanogrammus (Polynoidae)

Metabonellia haswelli (Bonellidae)

IDENTIFICATION TIPS

The most common marine annelids, polychaetes, have bodies that are obviously segmented. Bundles of chaetae protrude from the side of each segment (although a hand-lens or microscope may be necessary to see them). Other common kinds of marine worms—for example flatworms, ribbon worms, and peanut worms—have no visible segments or chaetae. Spoon worms (treated in most books as Phylum Echiura) do have chaetae and have been shown by recent research to belong within the Annelida, even though they have lost their segmentation.

DIVERSITY

At least 1,000 species of polychaetes are known in southern Australia, and many more species undoubtedly await description by taxonomists. Marine oligochaetes may be equally diverse but are poorly known in Australian waters. Marine leeches are probably much less diverse, but these also have been barely studied. Worldwide the Annelida comprise over 16,000 described species.

Polychaetes are common, widespread and diverse in most marine and estuarine environments (a very few species occur in freshwater streams or lagoons). Overseas, polychaetes may be actively cultivated to treat waste from sewage treatment plants and aquaculture ponds, but naturally occurring communities of polychaetes, and other invertebrates, are equally important in maintaining healthy marine environments. The great taxonomic diversity of annelids is reflected in their biology. Feeding modes and diets include predators with impressively toothed or venomous jaws, filter-feeders crowned with filaments, species which collect surface particles with ciliated tentacles and those which burrow to consume buried detritus. A few are parasites or commensals on other invertebrates. Most live on or near the seafloor, but a few are strong-swimming pelagic forms. Oligochaetes are burrowing deposit feeders. Marine leeches suck the blood of sharks and rays. Reproductive biology is similarly diverse and defies brief summary: sexual reproduction is normal but asexual modes occur across a variety of polychaetes. Sexes are often separate, but some

Pionosyllis sp (Syllidae)

Sabellastarte sp. (Sabellidae)

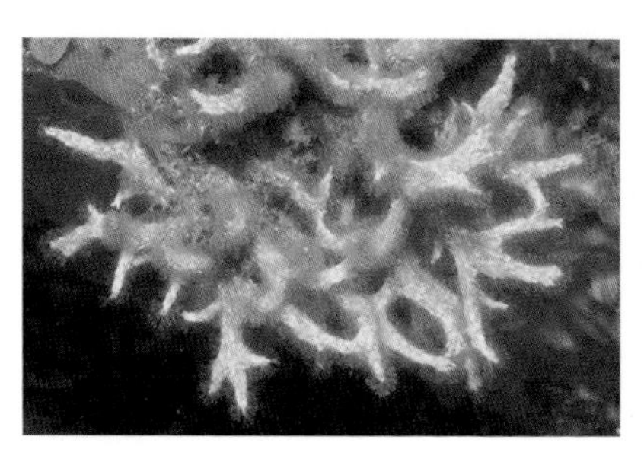

Salmacina australis (Serpulidae)

Unidentifed polychaete (Cirratulidae)

are hermaphrodites. Fertilisation may be external or internal, and larvae may develop directly into juveniles or there may be a planktonic larval phase.

Kamptozoans or nodding heads (Phylum Entoprocta)

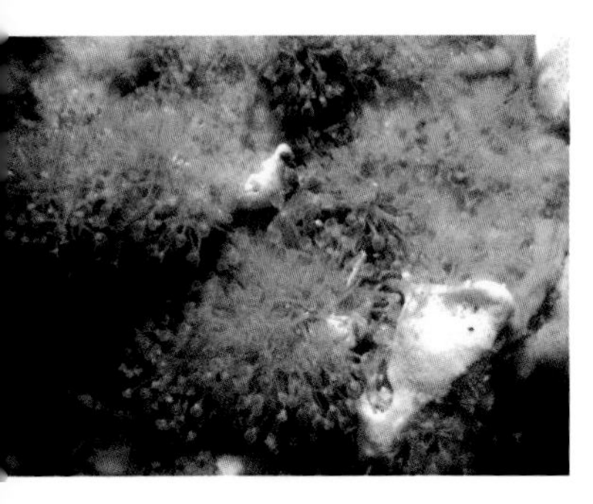

Pedicellina whiteleggi (Pedicellinidae)

Kamptozoans are tiny filter-feeding invertebrates that live attached to the sea-floor, or to larger sessile creatures such as bryozoans. They may be solitary, or may form colonies. Each kamptozoan individual (zooid) looks a little like a tiny stalked hydroid or anemone. Unlike hydroid polyps, however, the stalks of kamptozoans are muscular so that the little zooids have a characteristic bending movement leading to the delightfully descriptive common name 'nodding heads'. Kamptozoans are almost entirely marine (there is a single freshwater species) with about 150 species worldwide. A dozen or so species are known from southern Australia, several of which are yet to be given formal scientific names.

Horseshoe worms (Phylum Phoronida)

Phoronis hippocrepia (Phoronidae)

Horseshoe worms (Phylum Phoronida) are another phylum of worms which are not closely related to any of the other types of worm. Phoronids live in tubes where the bulbous rear of the body is hidden. The visible portion is a pair of horseshoe- or spiral-shaped feeding tentacles. Phoronids are not very diverse and are infrequently seen, but occur widely in the marine environment. Worldwide there are only about twelve species, four of which occur in southern Australian coastal waters.

Roundworms (Phylum Nematoda)

Nematodes, or roundworms (Phylum Nematoda) include both free-living and parasitic worms. Roundworms are highly abundant worms in many environments. There are a vast number of marine species, but they are mostly microscopic in size and usually only seen if special collecting techniques are used. Nematodes large enough to be seen in sediment samples are very smooth and shiny, and taper towards either end.

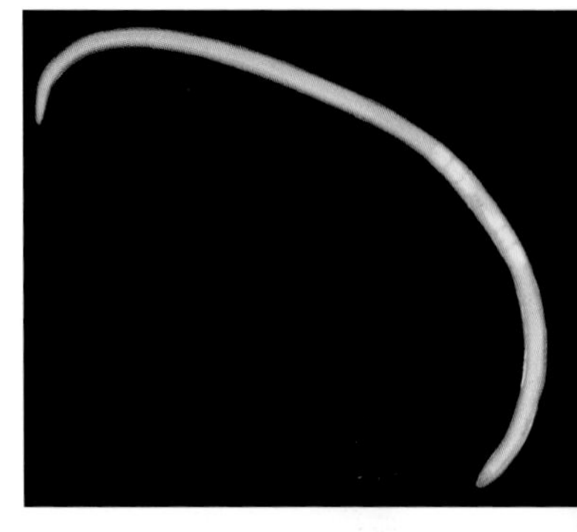

Unidentified nematode

Penis worms (Phylum Priapula)

Priapulans, suggestively known as penis worms, are unsegmented marine worms reaching 100 mm or more in length. One end has eversible mouth-parts and twenty or twenty-five longitudinal rows of hardened papillae, while the posterior region is the abdomen, with a terminal anus. There are numerous circular muscle bands around the abdomen. Priapulans are predators found in sand and mud on the sea floor. Only seventeen species of Priapula are known worldwide; they are sparsely distributed and uncommon. Little is known of penis worms in Australian seas although they are sighted from time to time.

Priapulus caudatus (Priapulidae)

Leptomithrax gaimardii (Majidae)

Arthropods
(Phylum Arthropoda)

Arthropods are characterised by their segmented body, jointed legs, and hard cuticle (exoskeleton) covering the body. The classification of arthropods is complex and evolutionary relationships in the phylum are still a matter of debate among scientists. Some primitive forms, such as fairy shrimps (Order Anostraca) and shield shrimps (Order Notostraca) live in inland saline waters and other even more obscure forms live in submerged caves or in the deep sea.

Most famous among primitive arthropods are the trilobites. Along with some other less well-known groups, trilobites became extinct about 250 million years ago (before the time of the dinosaurs) and are known only as fossils. One group of arthropods, apparently little changed for hundreds of millions of years, is still represented by a single living species the Horseshoe Crab, *Limulus polyphemus* (not found in Australian waters). Other groups of arthropods—principally insects and spiders on land, and crustaceans in the sea—radiated to become the most diverse and abundant of all animals. Arthropods are now thought to make up about 80% of all the known animal species.

In the marine environment, the major group of arthropods are the crustaceans. However, there are some marine insects and mites. Sea spiders (pycnogonids) only occur in the sea and despite the name, are not closely related to true spiders. This book includes most of the marine arthropod groups commonly found in southern Australia but some obscure and rarely seen kinds have had to be left out (see Further Information for references to others).

Insects (Phylum Arthropoda, Subphylum Hexapoda)

Although insects are the most diverse of all animals, the vast majority live on land or in fresh water. There are only a few truly marine insects. Those most likely to be found in the sea are the worm-like larvae of flies belonging to the family Chironomidae (Order Diptera). A chironomid larva usually has an obvious head which is hardened and dark brown. There is only one pair of antennae (not two as in Crustacea). There are usually three broad thoracic segments, followed by nine narrower abdominal segments. Paired fleshy, 'false legs' with claws occur on the first thoracic segment and the last abdominal segment. Hair-like setae also occur on the thorax and abdomen. There are numerous species of chironomids and they may be quite common in muddy sediments of estuaries.

Chironomus sp. (Chironomidae)

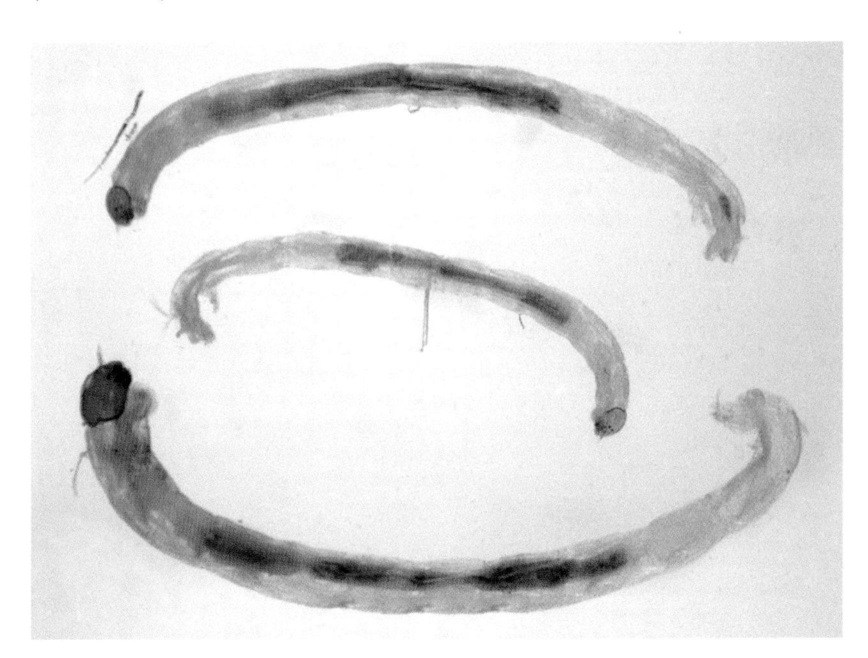

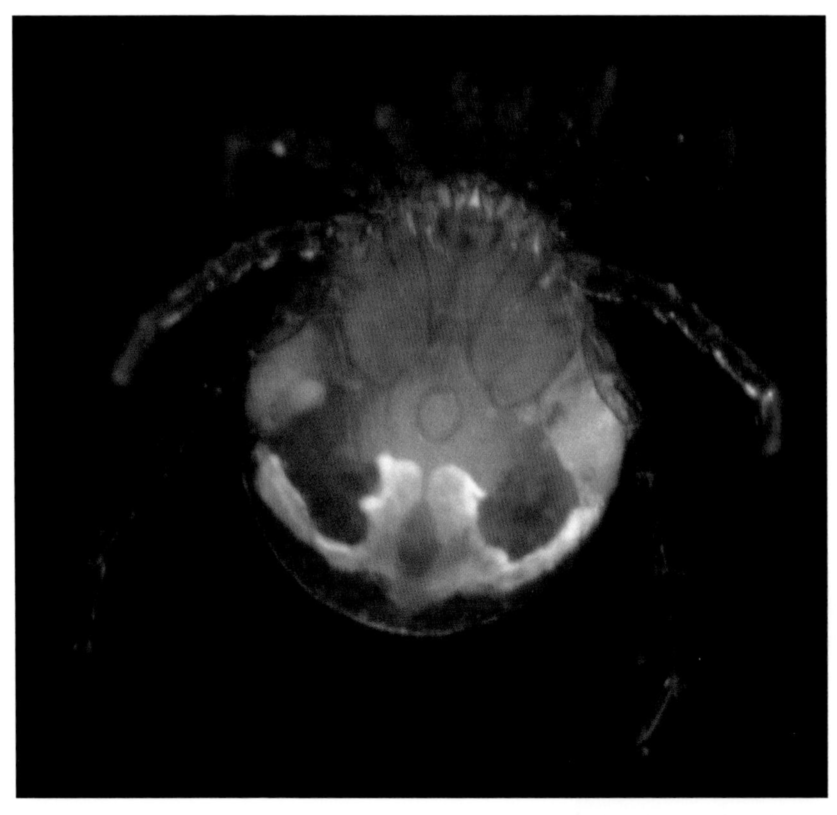

Halicarid sp. (Halicaridae)

Mites

(Phylum Arthropoda, Order Acari)

Hidden in the sand or attached to algae are the marine mites. These arthropods are more closely related to land-dwelling spiders than to crustaceans. They have four pairs of legs and no antennae, and most are less than 1 mm long. There are more than 30,000 known species of mites although most of these live on land. However, marine mites are common enough among the tiny animals that live on the seafloor and among seaweeds. Although minute, they are often bright orange or red, and quite conspicuous.

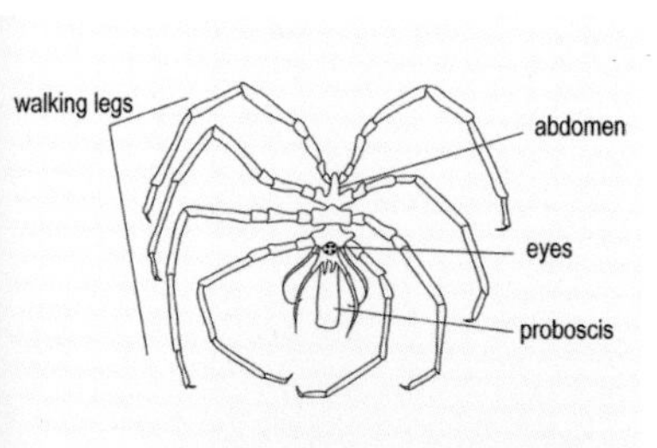

Generalised pycnogonid

Achelia sp. (Ammotheidae)

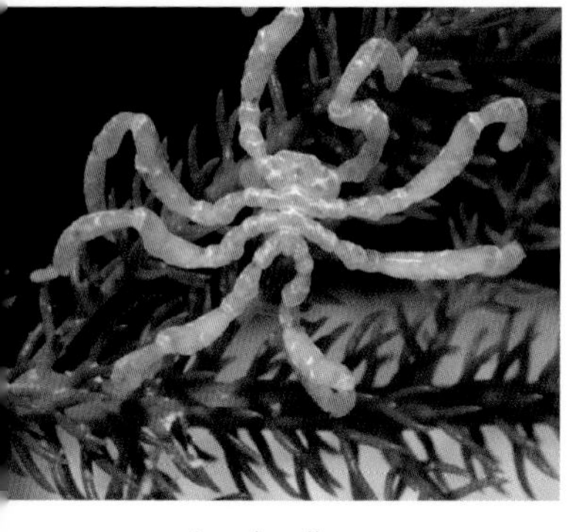

Pseudopallene sp. (Callipallenidae)

Sea spiders (Phylum Arthropoda, Subphylum Pycnogonida)

The long legs of the Pycnogonida earn them their common name: sea spiders. These marine animals have a small body with four, or rarely five or six, pairs of legs attached.

Most pycnogonids are small, with a 10–20 mm leg span, but some deeper sea species reach 500 mm. At the front of the body is the proboscis, with the mouth at the tip. Behind the proboscis is the trunk, with a tubercle bearing typically four, or as many as eight, pairs of eyes. The walking legs are attached to the sides of the trunk. The abdomen is reduced to a single segment. A cuticle covers the body, which may be smooth or bumpy, bare or with setae or spines, sensitive to touch and chemicals.

In addition to walking legs, the appendages include palps, chelifores and ovigers. The palps are sensory while the chelifores gather food. The ovigers hanging below the body are for cleaning and, in the male, for holding the eggs.

Pycnogonids occur world-wide, including in the polar regions, from shallow waters to almost 7000 m deep. Although quite common in southern Australia, they are often overlooked because they are small and cryptic. Pycnogonids tend to crawl slowly over the sea floor, but may also swim or drift attached to hydroids and seaweeds. If you look carefully you might see a sea spider hiding among hydroids or bryozoans, floating along the sea floor or in a rock pool.

IDENTIFICATION TIPS

Pycnogonid species are identified principally by the number and structure of the appendages; other useful features include the shape and size of the proboscis, relative lengths of body and legs, and body shape. Pycnogonids may be transparent or coloured. Some species which live on hydroids have the same colour as their host, such as cream or yellow. Those of the deep sea are often orange to red, as these colours are invisible at depth.

DIVERSITY

There are more than 1300 pycnogonid species world-wide. More than 50 species are described from southern Australian shallow waters, with many more known from the deep waters of the southern ocean.

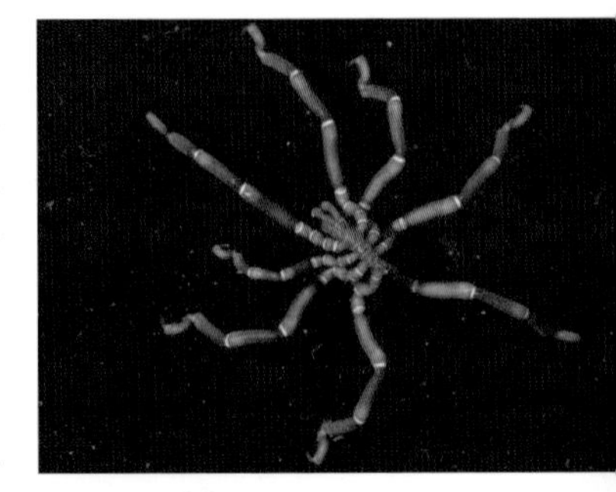

Anoplodactylus evansi (Phoxichiliidae)

ECOLOGY

Pycnogonids live in all seas and at all depths. The proboscis is specialised to feed on the body fluids of sessile animals such as sponges, anemones, hydroids or bryozoans. Others prey on mobile forms like polychaete worms. It is believed that most species are parasitic at some stage of their life. Pycnogonids have separate sexes. After mating, the male gathers the eggs laid by the female on to its ovigers where the embryos develop. The larva that is released often lives as a parasite on another invertebrate.

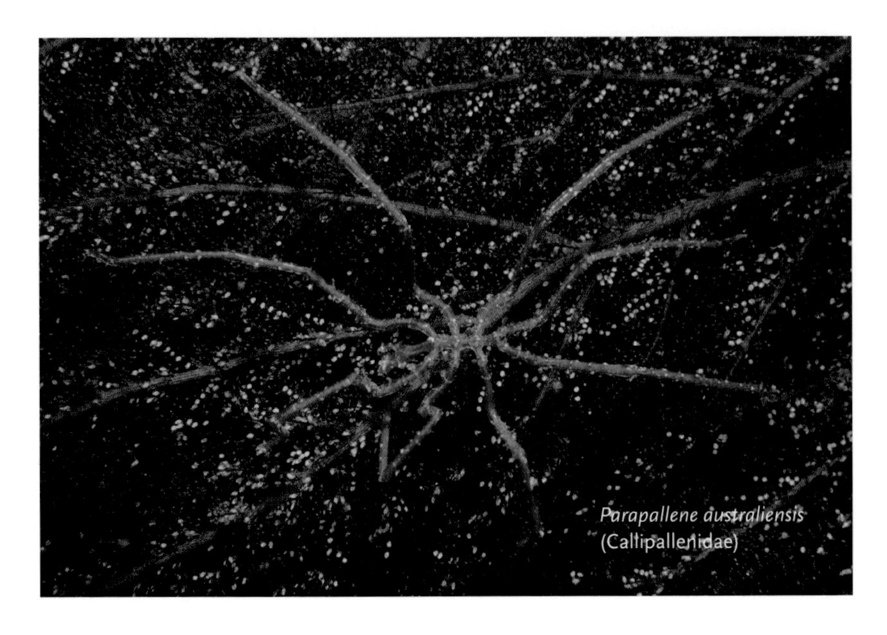

Parapallene australiensis (Callipallenidae)

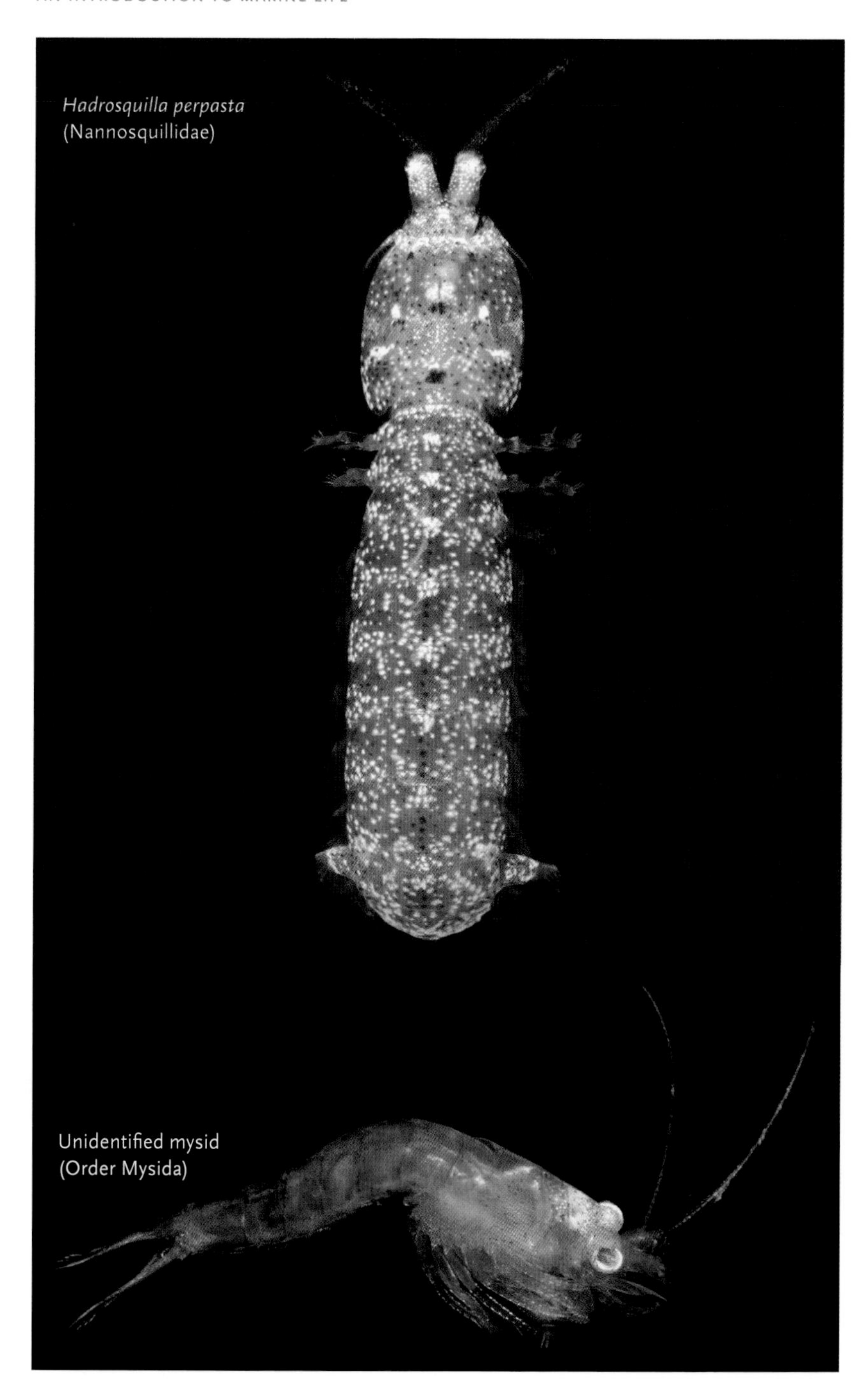

Hadrosquilla perpasta
(Nannosquillidae)

Unidentified mysid
(Order Mysida)

Crustaceans (Phylum Arthropoda, Subphylum Crustacea)

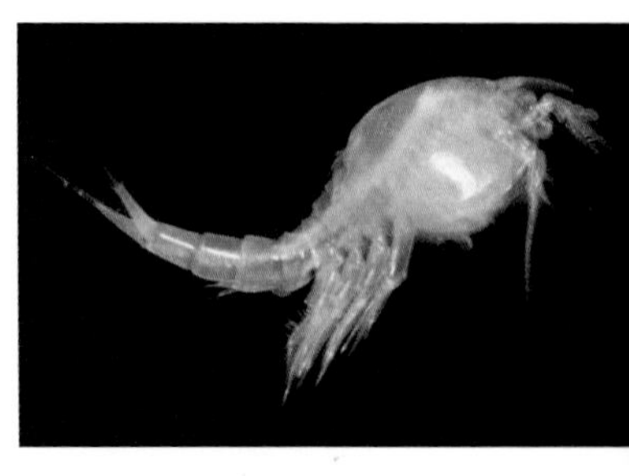

Nebalia sp. (Nebaliidae)

Crustacea differ from other Arthropoda by having two pairs of antennae on the head, although these may be lost in the adult forms. Some live on land or in fresh water, but most are marine. Crustacea are often thought of as the 'insects of the sea'—filling every niche, and specialised for many modes of life.

The body of the crustacean is divided into three sections: head, thorax (or trunk) and abdomen. These sections may be difficult to distinguish, especially if the body is covered by a carapace. A carapace, or 'shell' is an extension of the cuticle over some or all of the body segments and their legs and gills. The body has many pairs of appendages—these are usually specialised for particular functions. At the front are feeding appendages, followed by legs for walking or swimming, and perhaps appendages modified into gills or for holding young.

Reproduction in crustacea is diverse. The sexes are typically separate but during growth many crustaceans change from male to female, or vice versa. During mating, males usually inseminate females using specially modified penis-like limbs. Developing eggs and larvae may be planktonic with a potential for very wide dispersal, or may develop directly attached to, or brooded by the adults.

More than 68,000 crustacean species have been described. About half of all crustaceans belong to the Class Malacostraca, which includes the decapods (crabs and lobsters), isopods, amphipods, tanaids, cumaceans, mantis shrimps, krill, nebaliaceans, mysids, and other less familiar forms. This class is defined by the arrangement of segments in different parts of the body: five segments in the head, eight in the thorax and six in the abdomen. There are five other classes of crustacea but these include mostly obscure types that are either confined to fresh water, or are tiny or parasitic. The only familiar exceptions are ostracods (seed shrimps), copepods, and barnacles (see p70 and 78).

Unidentified euphausiid (Euphasiidae)

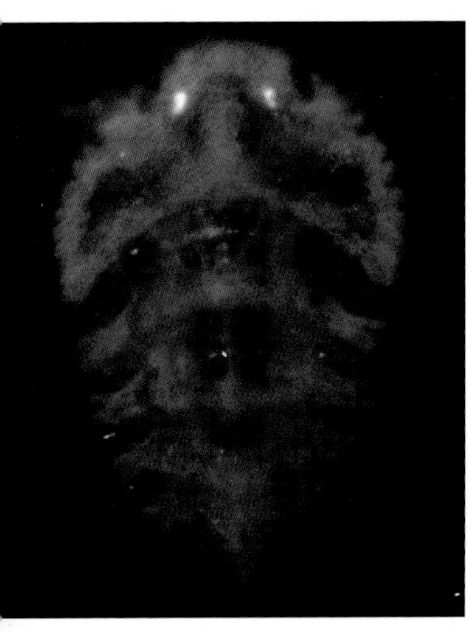

Unidentified copepod
(Order Harpacticoida)

Copepods (Subphylum Crustacea, Subclass Copepoda)

Copepods are minute crustaceans, not often seen yet they occur in huge numbers, and are highly diverse. There are over 200 families of copepods and a vast number of species. Without a powerful microscope, most features of a copepod cannot be seen. But a distinctive feature is the cephalon (a shield-like carapace) with a single simple eye; subsequent segments becoming narrower posteriorly. There are a number of minute limbs and other appendages. Female copepods are most easily recognisable if they are trailing a pair of conspicuous egg sacs. Copepods are an especially important component of the plankton but most are so small that they can pass straight through a net with the openings larger than quarter of a millimetre. Other copepods live on the sea floor, and many are parasites of fish and other marine animals.

Unidentified ostracod
(Cylindroleberididae)

Ostracods or seed shrimps (Subphylum Crustacea, Class Ostracoda)

Ostracods have a distinctive enlarged carapace in two parts that encloses the entire body. They hardly look like crustaceans at all, and might initially be mistaken for tiny bivalve molluscs. Jointed limbs protruding from the carapace (when it's open) are a giveaway that they are crustaceans. Within the carapace the trunk is reduced; there are up to 8 pairs of limbs. There are many hundreds of species of ostracods in southern Australia, and many of these are undescribed. Few shallow water species exceed 5 mm in size, and most are much smaller. Many ostracods are scavengers and scurry quickly over the sea floor. Some species are strong swimmers including some that are bioluminescent (these are one of several animal groups that cause bright flecks of bioluminescent light in the sea at night).

Tanaids (Subphylum Crustacea, Order Tanaidacea)

Apseudes sp. (Apseudidae)

Tanaids are small crustaceans that are very common in shallow water environments yet are unfamiliar to most marine naturalists. They are roughly similar to isopods and the two groups can easily be confused. Tanaids can be distinguished by their large chelipeds (nippers). Tanaids are mostly cylindrical in form, and their head is fused to the first two thoracic segments forming a short carapace. The carapace is followed by a thoracic region with five pairs of legs. At the posterior end, the tail-end limbs (uropods) have long thin appendages looking a little like antennae. Most tanaids in Australian environments are smaller than 10 mm. They are found in many habitats including soft sediments and among algae. Unlike most other crustaceans, many tanaids are hermaphrodites; they may start out as either male or female and later change to their opposite sex.

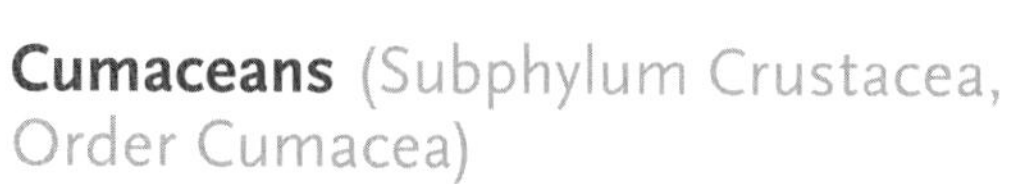

Cumaceans (Subphylum Crustacea, Order Cumacea)

Cyclaspis usitata (Bodotriidae)

Cumaceans are another kind of small crustacean that, despite being common, are not well known. Cumaceans are distinctive and fairly uniform, having an enlarged carapace which is fused with some thoracic segments. The abdomen is narrow and terminates with a pair of rod-like uropods. Most common Australian cumaceans are less than 10 mm long. Cumaceans are often abundant in shallow waters, especially in sand and mud sediments.

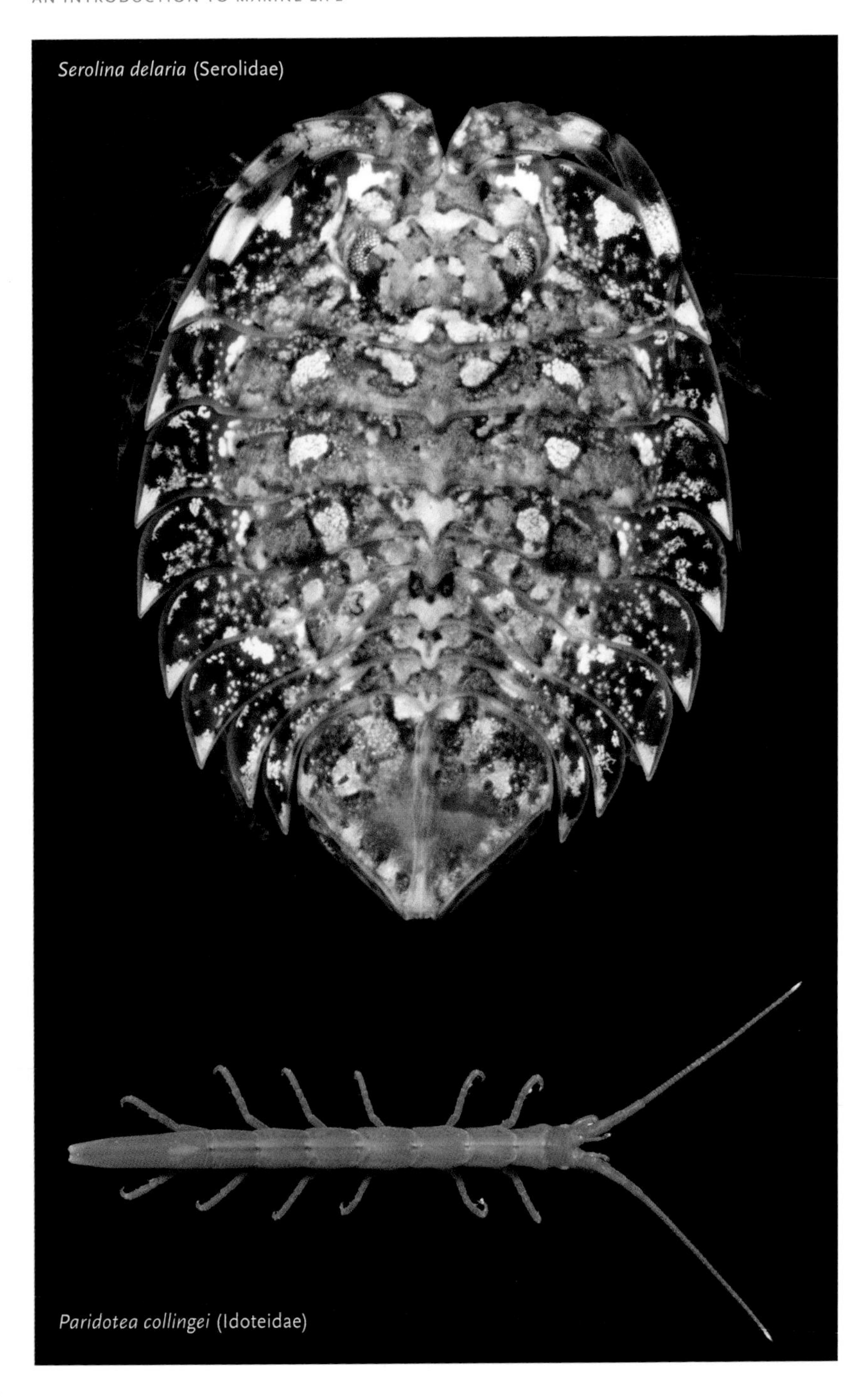

Serolina delaria (Serolidae)

Paridotea collingei (Idoteidae)

Pill bugs, sea lice
(Subphylum Crustacea, Order Isopoda)

Isopods are small crustaceans also known as sea lice, pill bugs and beach slaters. Most are about 5 mm long, but they can range in length from 0.5 mm to 500 mm. Isopods are found in marine, freshwater and terrestrial habitats—the common garden slater is an isopod.

IDENTIFICATION TIPS

Isopods have one segment of the thorax fused to the head. They have seven pairs of legs and one pair of tail-end limbs (uropods) at the end of the abdomen. There is no typical body form; isopods may be flattened, long and thin, or compact.

DIVERSITY

More than 10,000 isopod species have been described, in 95 families. Over 1,100 species are known to occur in Australian waters. Isopods are one of the specialised groups that have become morphologically diverse and rich in species in the deep sea.

ECOLOGY AND HABITAT

Marine isopods are specialised for numerous habitats and food sources. Most species live on the sea floor at depths ranging from intertidal rock platforms to the deep sea. One family of deep-sea isopods has been recently discovered to include swimming species that never touch the bottom. Particular microhabitats include the fronds of seaweeds and the inside of sponges. Food sources are dead fish, algae, and the cellulose from wood. Some isopod species are notorious parasites of fish. They are commonly known as 'fish lice' and live attached to the gills of a fish, the body or inside its mouth.

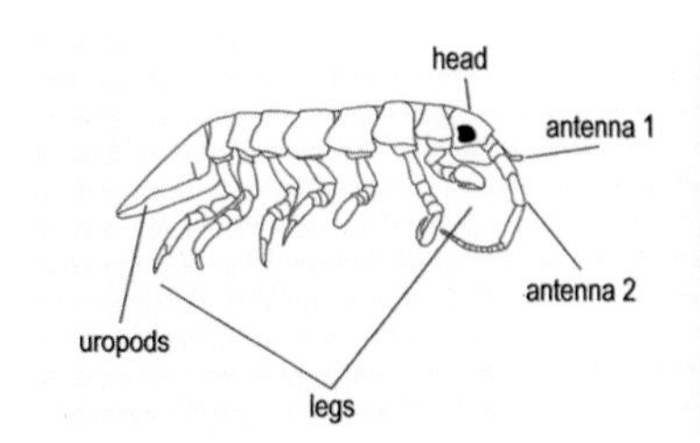

Generalised isopod

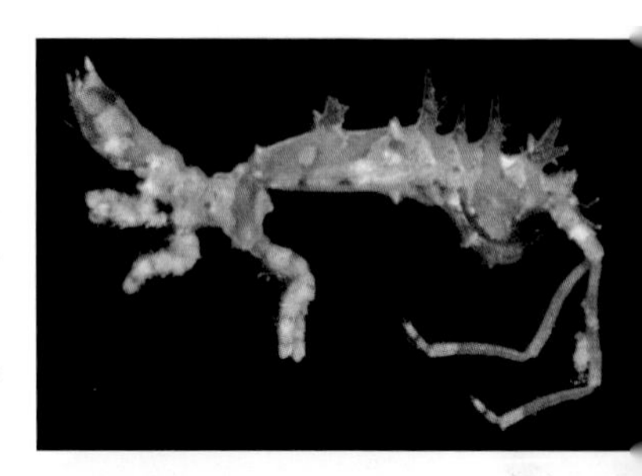

Neastacilla barnardi (Arcturidae)

Ligia australiensis (Ligiidae)

Sphaeromatid sp. (Sphaeromatidae)

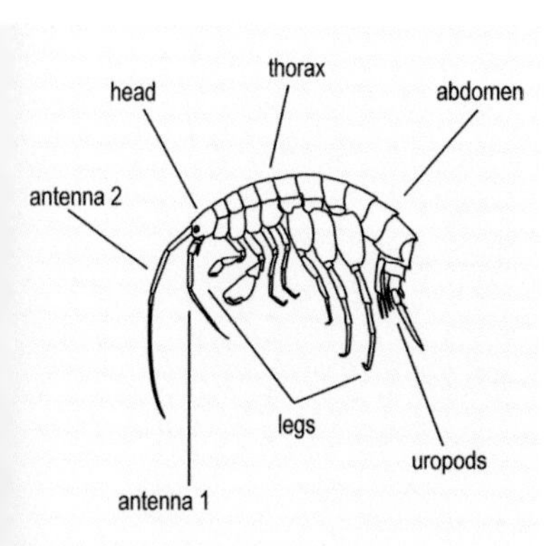

Generalised amphipod

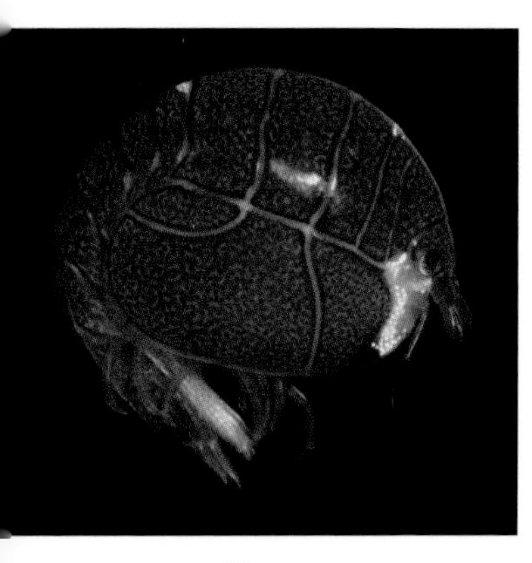

Cyproidea marmorata (Cyproideidae)

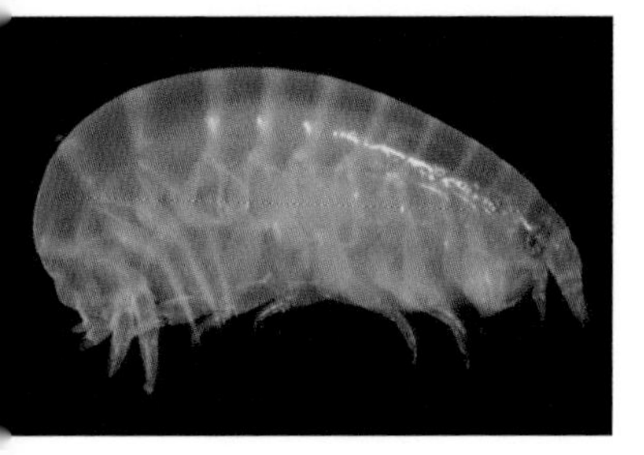

Parawaldeckia sp. (Lysianassidae)

Amphipods (Subphylum Crustacea, Order Amphipoda)

Amphipods are common components of the rocky shore and shallow marine environment. Their habitat extends to the deep sea, where some species grow to 300 mm long. The typical length is 3 to 6 mm. Some amphipod species jump about under clumps of seaweed on the shore, giving them their common name of sand hoppers.

IDENTIFICATION TIPS

Amphipods share several features with isopods: one segment of the thorax is fused to the head and there are seven pairs of legs. They can be distinguished from isopods by having three pairs of limbs (uropods) at the tail end, pointing backwards.

There are two major body types in the Amphipoda. The Gammaridea have a compact body, flattened from side to side and are able to swim and scuttle rapidly over the reef and sea floor. The Caprellidea are long and angular, like 'stick figures', and move much more slowly on clasping legs.

DIVERSITY

There are more than 8,000 described species of amphipods, divided into almost 200 families. There are about 1200 species known from Australian waters.

ECOLOGY AND HABITAT

The many different species are specialised for particular habitats, including some on land. Marine amphipods live at all latitudes. Most are benthic, living on the sandy and muddy sea floor, on reefs, amongst seaweeds and bryozoans, and under rocks on rocky intertidal shores. Some species are wood borers and can be found burrowing into timber structures such as piers. Many amphipods construct tubes out of mud or of secreted silk.

Most amphipods are scavengers of dead plant and animal matter. Some eat living algae, bacteria in the sediment and particles filtered from the water. Members of one family have legs with claws that enable them to attach to dolphins and whales.

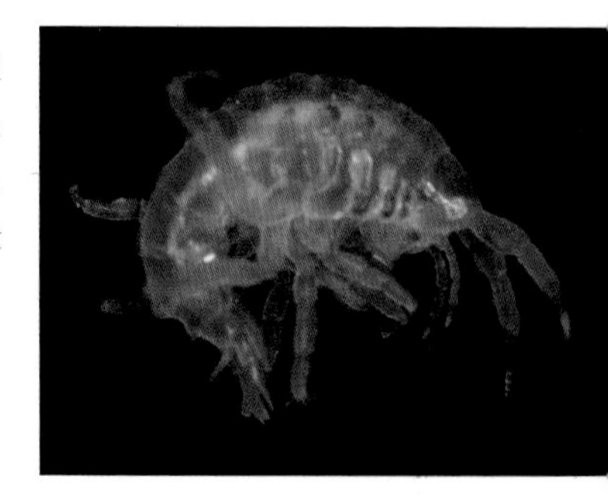

Syndexamine mullauna (Dexaminidae)

Unidentified amphipod (Melitidae)

Bellorchestia richardsoni (Talitridae)

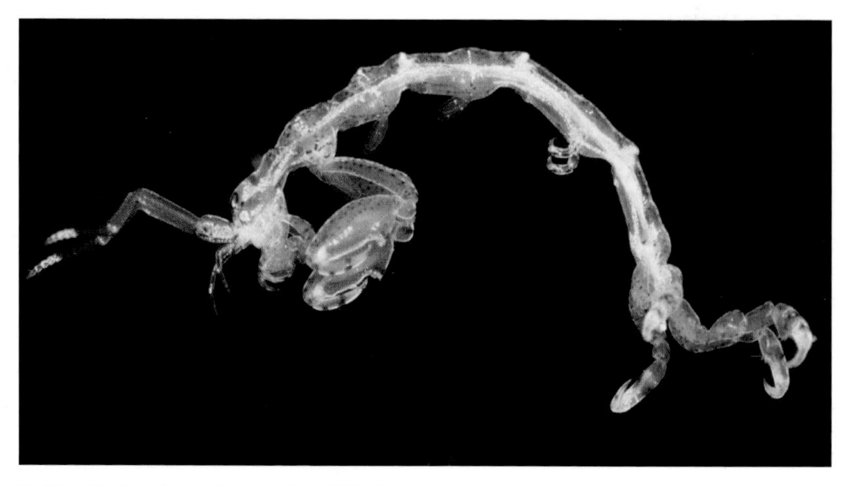

Unidentified skeleton shrimp (Caprellidae)

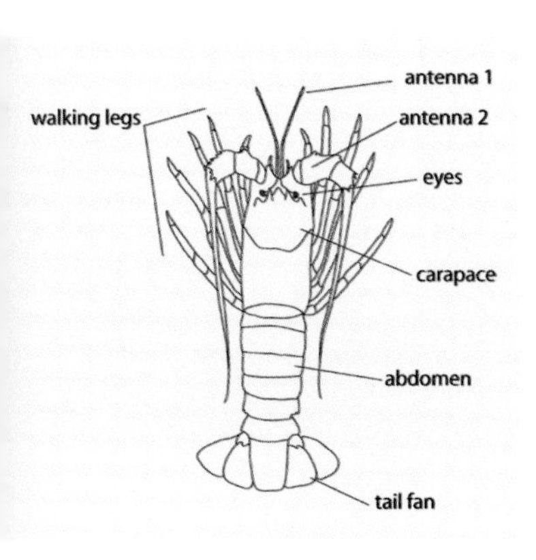

Generalised decapod

Unidentified snapping shrimp (Alpheidae)

Cyclograpsus sp. (Grapsidae)

Crabs, shrimp, lobsters (Subphylum Crustacea, Order Decapoda)

The largest and most familiar crustaceans are the crabs, prawns, shrimps and lobsters, which are classified within the Decapoda. These animals are named for their ten legs. In many examples the first pair of legs are modified as claws (chelipeds). Decapods have three small pairs of thoracic appendages modified for feeding, in addition to those on the head.

Crabs belong in a group called Brachyura, defined by the possession of one pair of claws, four pairs of walking legs, and the short abdomen (tail) tucked under the body. There are many different types of brachyuran crabs, some conspicuous on the rocky intertidal shore or seen scuttling across the sea floor. Shrimps, or Caridea, have a long muscular abdomen and usually two pairs of claws. The snapping shrimp is a typical example. The Anomura are another group of decapods having the last pair of walking legs much smaller than the other legs. This group includes squat lobsters with long claws and hermit crabs which have a soft coiled abdomen and usually live in empty gastropod shells. Australia's marine rock lobsters (popularly called *crays*) and Balmain bugs belong in the Achelata. They have no claws, a thick carapace over the head and thorax, a long abdomen and a tail fan. Rock lobsters especially are an important commercial fishery. Prawns, another important edible crustacean, belong in the decapod group Dendrobranchiata which unlike most others, do not carry and incubate their eggs.

IDENTIFICATION TIPS

The different groups of decapods can be distinguished by the arrangement of legs, how many claws are present, and the size of the abdomen. To identify individual species of decapod, look at body shape and size, ornamentation of spines and teeth, carapace colour and pattern and structure of the appendages.

DIVERSITY

All of the major decapod groups are represented in southern Australia. In waters less than 100 m deep, almost 400 species are recorded. Half are brachyuran crabs.

ECOLOGY AND HABITAT

Decapods live in rocky intertidal platforms, sandy shores, estuaries, under water reefs, shallow seas and the deep sea floor. As they are eaten by fish, marine mammals and humans, many species have found ways to hide or disguise themselves. Some burrow into the sediment on the sea floor, others prefer to hide in crevices, and hermit crabs protect themselves with empty gastropod shells. The decorator crab cuts pieces of seaweed, sponge and hydroid and attaches them to hooks on its carapace. Many crabs have impressive claws for defence.

Decapods are usually predators or scavengers, eating echinoderms, bivalves, polychaetes, and crustaceans. Some decapods are filter-feeders, using fringed-appendages to extract food particles from the water. A few decapods are herbivores, eating seagrasses and seaweeds.

Galathea australiensis (Galatheidae)

Ibacus peroni (Scyllaridae)

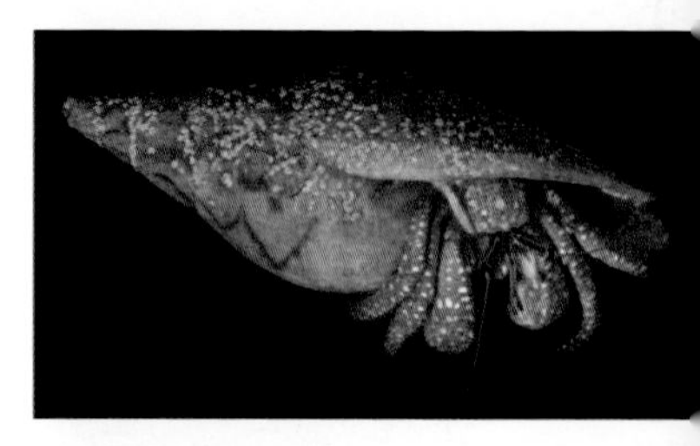

Strigopagurus elongatus (Diogenidae)

Upogebia simsoni (Upogebiidae)

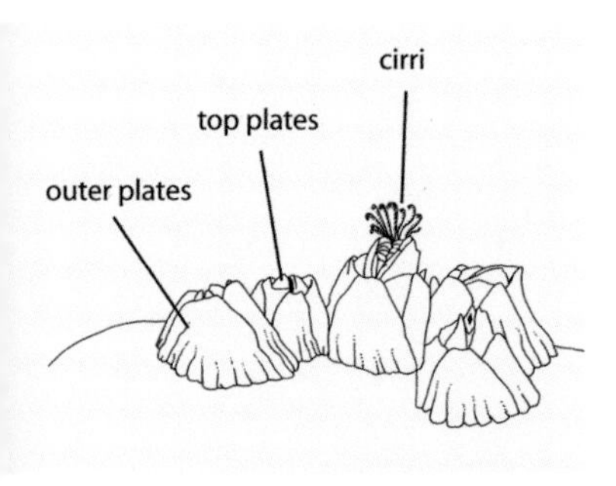

Generalised barnacles

Chamaesipho tasmanica (Chthamalidae)

Austromegabalanus nigrescens (Balanidae)

Barnacles (Subphylum Crustacea, Class Cirripedia)

Barnacles are often seen on intertidal rocky platforms, pier pilings and the hulls of ships. Unlike other crustaceans, they are attached to the surface and cannot move around as adults. Their body is surrounded by hard plates and their segmented legs (cirri) extend through the top opening to catch food. The common barnacles fall into two groups. Acorn barnacles have no stalk and the plates are in contact with the surface. Goose barnacles attach by a long stalk and the plates surround the separate body. Goose barnacles usually live in the open sea attached to floating objects and are seen only when they wash ashore. Some parasites related to barnacle infest other invertebrates, crustaceans in particular.

IDENTIFICATION TIPS

At first glance, barnacles resemble molluscs such as limpets on the rocky shore platform. However, a barnacle has a series of outer plates and top plates instead of one continuous shell. To distinguish different barnacle species, look at the presence or absence of a stalk, the number and arrangement of the shell plates, ornamentation of the shell plates (such as ribs or stripes), shape (tall cone or flattened), size and habitat.

DIVERSITY

Southeastern Australia is home to 80 barnacle species. Although the intertidal and shallow water fauna is well documented, the smaller parasitic species and those of the deep sea are less well known.

ECOLOGY AND HABITAT

Barnacles release fertilised eggs into the plankton, where they drift and develop through several stages before settling onto a site and attaching. The most common place to find barnacles is on the rocky intertidal shore platform. Species are adapted for life at different levels on the shore, some preferring the high, drier zones while others live at the edge of the platform. Some species live on the exposed surface while others live in shady crevices. Other barnacle species live in estuaries, on piers and on the hulls of ships where they may be transported around the world.

Adult barnacles may be eaten by gastropod molluscs and sea birds. Planktonic barnacle larvae are eaten by fish and many invertebrates, including adult barnacles.

Ibla quadrivalvis (Iblidae)

Catomerus polymerus (Catophragmidae)

Lepas anserifera (Lepadidae)

Tetraclitella purpurascens (smaller species is *Chthamalus antennatus*) (Tetraclitidae)

Ceratosoma brevicaudatum (Chromodoridae)

Molluscs (Phylum Mollusca)

The Phylum Mollusca are animals with a head, a muscular 'foot' and the gut cavity covered by a fleshy layer known as the mantle. The mantle secretes a shell (in many molluscs) and also protects the gills. This basic body plan can be found, with variations, in all molluscs. The most familiar molluscs inhabit a wide range of marine environments: the bivalves—mussels, clams and relatives (Class Bivalvia); the snails and slugs (Class Gastropoda); the squid, cuttlefish and octopus (Class Cephalopoda); and the chitons (Class Polyplacophora). Tusk shells (Class Scaphopoda) are less familiar molluscs with a tapered, slightly bent shell. Class Aplacophora are obscure worm-like small molluscs rarely seen in shallow water. Class Monoplacophora are primitive limpet-like molluscs found only in the deep sea, and not yet known from Australia.

As with the other large phyla, reproduction in molluscs encompasses a great diversity of strategies. Sexes are separate in most bivalves, chitons and cephalopods. Copulation occurs in a few gastropods and most cephalopods. Many cephalopods grow quickly then reproduce and die, but most other molluscs live on to reproduce in successive years. Gastropods may be hermaphrodites or may have separate sexes (they may be simultaneously hermaphroditic, or may change from male to female during their life). After fertilisation, the developing larvae may be planktonic or they may be retained and brooded by the adults.

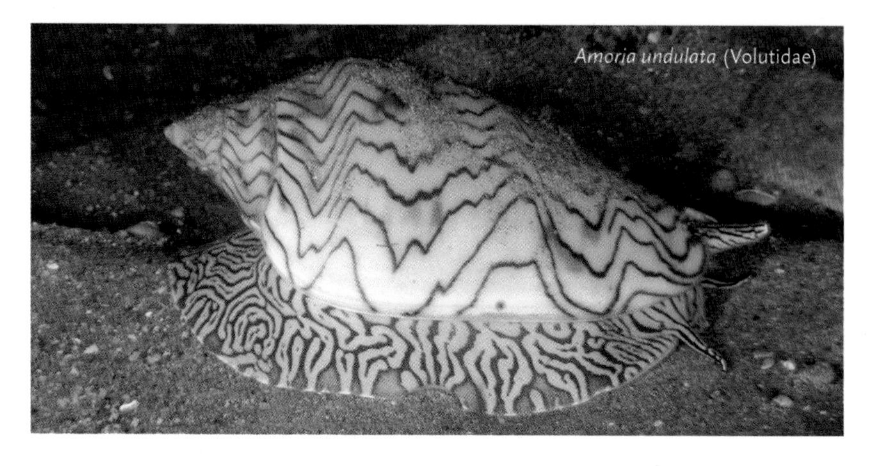
Amoria undulata (Volutidae)

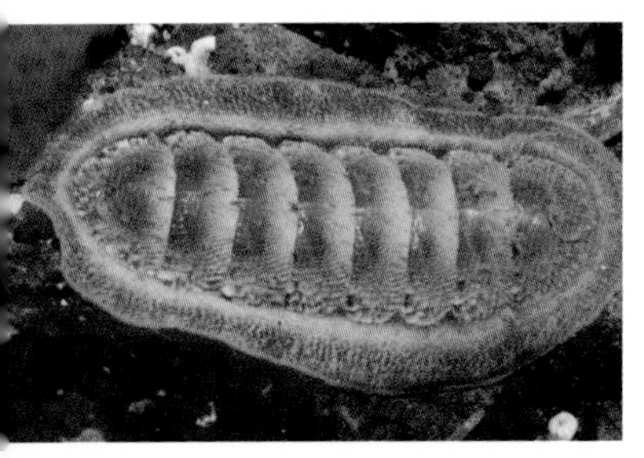

Ischnochiton cariosus (Ischnochitonidae)

Ischnochiton elongatus (Ischnochitonidae)

Cryptoplax striata (Acanthochitonidae)

Chitons (Phylum Mollusca, Class Polyplacophora)

Chitons are easily recognised by their flattened elongate body with eight separate shell plates. They are most commonly found on intertidal rock platforms under rocks and crevices in the surge zone.

IDENTIFICATION TIPS

Chitons are easily identified by their eight wide plates. In some groups the plates are greatly reduced and the animal is more elongate or cigar-shaped. Around the edges of the shell plates is a toughened area known as the girdle. It often bears calcareous or chitinous spines, scales or bristles. There are no eyes or tentacles. Instead some species have eye-like structures in pits in the shell plates. A toothed tongue (radula) is present. The grooves, lines or ridges on the shells help with species identification. Many species have attachment teeth on the valves for locking with the adjacent valve.

DIVERSITY

There are at least 1,000 species known in the world, and 171 species are recorded from Australian waters. Chitons are exclusively marine and most typically found in intertidal areas. They also occur at all depths to the deep sea.

ECOLOGY

Chitons typically feed on algae and small animals that they scrape off the rocks using their toothed tongue. They are more active at night, slowly moving over rocks as they graze. Some can defend themselves by curling up in a ball to protect the vulnerable muscular foot. Some species brood their young in the groove between their foot and the outer girdle.

Tusk shells

(Phylum Mollusca, Class Scaphopoda)

One small, uncommon group of molluscs is the distinctive tusk shells. These animals live buried in soft sand and mud. The shell is usually 50 mm or less and is a single hollow tube, typically tapered. The animal lives head down with the narrow end just above the sediment surface to draw water in for the gills. Tusk shells feed on microscopic organisms in the sediment such as forams, using retractable clubbed tentacles. There are about 900 species worldwide, and about 100 in Australian waters, most occurring in deep water. The most common shallow-water genus, *Cadulus,* has a shell that bulges in the midsection.

Dentalium intercalatum (Dentaliidae)

Aplacophorans

(Phylum Mollusca, Class Aplacophora)

Another small, rarely found group of worm-like molluscs is the aplacophorans. They are so uncommon that they do not even have a common name. Aplacophorans are mostly small (around 5 mm long) and covered with a glistening felt of calcareous spicules, although these spicules are not always obvious. Some aplacophorans (the Neomeniomorpha) have a ventral 'foot' which is similar to the foot of a snail or slug. Other aplacophorans (the Chaetodermomorpha), lack a ventral foot but in these the felt of spicules is dense and obvious. The continental shelf of southern Australia is one of the richest areas known for these obscure molluscs, with at least thirty-two aplacophoran species currently known from Bass Strait alone.

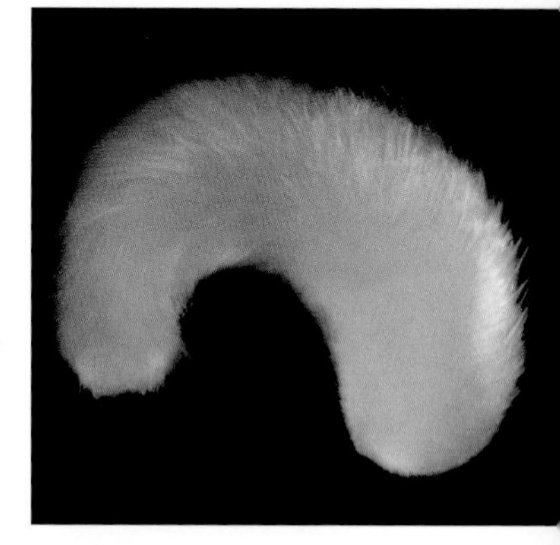

Eleutheromenia minimus (Pruvotinidae)

Austrocochlea spp. (Trochidae) and *Nerita* sp. (Neritidae)

Cellana tramoserica (Nacellidae)

Haliotis rubra (Haliotidae)

Snails, slugs
(Phylum Mollusca, Class Gastropoda)

The gastropods are a group of molluscs that are best known by the animals that make the familiar spiral seashells, but it includes all snails and slugs. Most species with a spiral shell can withdraw their body within the shell and close the entrance with a trapdoor (operculum). The body is asymmetrical. The spiral shell in many species is reduced to a wide hat (as in abalone or limpets) or a thin hidden shell (as in sea hares). There are many shell-less forms with the most dramatic group being the colourful sea slugs known as nudibranchs.

IDENTIFICATION TIPS

The features common to most gastropods are a single shell (in shelled forms), a head with eyes and/or sensory tentacles, and a muscular foot on which the animal travels. Most possess a toothed tongue (radula) that is used for scraping or drilling their food. Sea slugs tend to be oval to elongate, some with obvious external gills.

DIVERSITY

There are at least 70,000 species of gastropods in the world, found in all marine, freshwater and terrestrial habitats. There are three main groups: prosobranchs (most shelled marine snails), opisthobranchs (sea slugs) and pulmonates (air-breathing terrestrial snails and slugs).

ECOLOGY

Gastropods occupy a wide range of ecological niches from herbivores to top-level predators. Many species including those commonly found on intertidal rock platforms are grazers, using their toothed tongue to feed on algae and microscopic plant films. Others, such as dog winkles, are predatory, using the toothed tongue in a different way to drill through the sides of other shelled molluscs and poison the occupant. During development, gastropods rotate their body and organs 90 to 180 degrees against the foot. This is known as torsion and, in shelled forms, results in the head and anus being close together at the opening of the shell. Some gastropods change sex depending on the ratio of the different sexes. In a few groups, individual animals can possess organs of both sexes (hermaphrodites).

Dicathais orbita (Muricidae)

Austrolittorina unifasciata (Littorinidae)

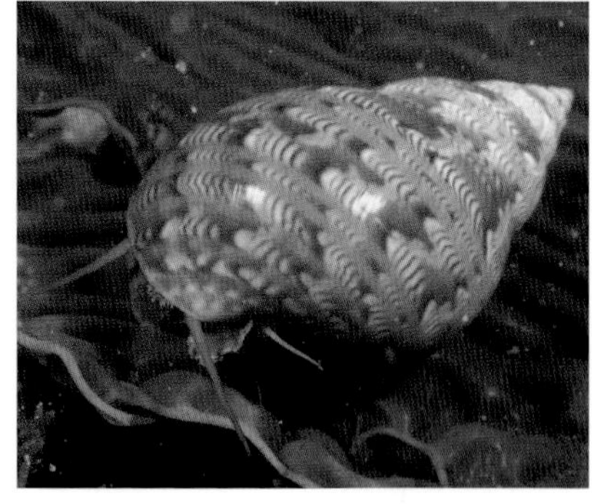

Phasianella australis (Phasianellidae)

Scutus antipodes (Fissurellidae)

Chromodoris tasmaniensis (Chromodorididae)

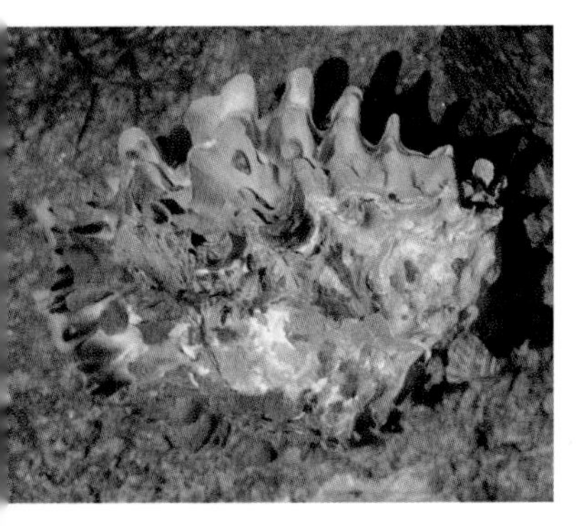

Crassostrea gigas
(Ostraeidae)

Circomphalus disjecta
(Veneridae)

Brachidontes rostratus
(Mytilidae)

Mussels, clams
(Phylum Mollusca, Class Bivalvia)

The bivalves are molluscs that have a two part shell, typically joined by a hinge. This group includes many popular seafood species. Shell shapes vary enormously from smooth round matching shells (as in many clams and cockles), to species with one curved and one flat shell (as in scallops), to species where one half of the shell (one of the 'valves') is glued to a rock (as in oysters), to highly modified burrowing species with reduced valves (such as the teredo or shipworm).

IDENTIFICATION TIPS

Most bivalves are easily recognised by their two shells joined together by a hinge. The shells can be calcified or horny, and can be closed tightly together using powerful muscles. Internally they have plate-like gills used for filtering food particles and planktonic organisms out of the water. There is no obvious head with eyes but some species (such as scallops) have a row of eyes along the gill fringes. Many species have a muscular foot that can be extended when burrowing to draw the shell down into sand or mud. Species such as mussels use strong hairs (byssal threads) to attach to rock, weed and wood.

DIVERSITY

At least 20,000 species occur in the world in both marine and fresh waters. These animals can be abundant but not obvious as many species remain buried and only extend the tips of their siphons to feed.

ECOLOGY

Most bivalves are filter feeders, setting up a current of water that runs over their gills. Food particles and plankton are trapped on mucous sheets that travel to the mouth, like a conveyor belt. The hardened shell offers protection, with strong muscles quickly closing the shell. Some snails and octopuses feed on bivalves by drilling through the shell with a toothed tongue and poisoning the occupant. Beach-washed shells often have holes some people use to thread on a string. These are drill holes made by the predator that killed them.

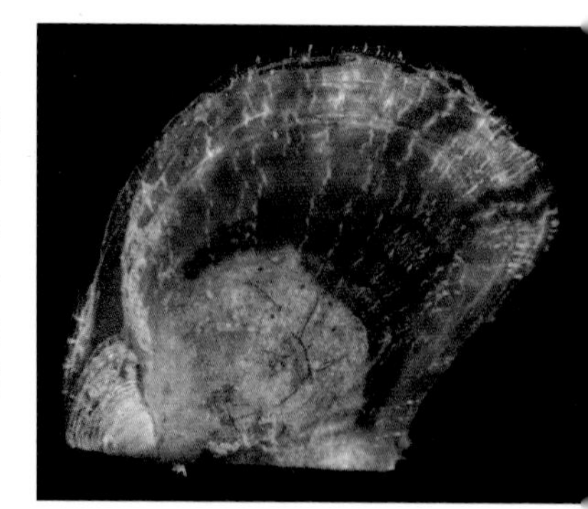

Electroma georgiana (Pteriidae)

Pecten fumatus (Pectinidae)

Atrina tasmanica

Mimachlamys asperrima (Pectinidae)

Argonauta nodosus (Argonautidae)

Cephalopods

(Phylum Mollusca, Class Cephalopoda)

Octopuses, cuttlefishes, squids and nautiluses are collectively known as cephalopods. The name comes from the Greek (kephale = head, podos = foot), meaning 'head-footed'. This refers to the ring of limbs attached directly to the head. Although popular as seafood, few people observe live cephalopods in the wild as they are expert at avoiding detection and capture.

IDENTIFICATION TIPS

All cephalopods have limbs that form a ring around the mouth. They have well developed eyes and brain and possess a horny two-part beak. All the body organs are held in a tubular or round sac at the posterior end, known as the 'mantle'. The mantle has a cavity on the underside in which the gills hang. Cephalopods obtain oxygen by drawing water into this cavity. The water is expelled through a tubular funnel ('siphon'), enabling jet propulsion.

Southern Calamari Squid, *Sepioteuthis australis* (Loliginidae)

The different cephalopod groups vary in other structures. The chambered or pearly nautiluses have a rigid shell where the animal occupies the largest outer chamber. None occur in southern Australia. Cuttlefishes possess a chalky internal cuttlebone. Many squid possess a transparent internal shell ('pen' or 'quill'). Most octopuses possess two tiny rod-like structures ('stylets') as remnants of a shell. Some squid and octopus completely lack a shell. The female argonaut is an octopus that produces and lives in a brittle white shell ('paper nautilus') that also functions as an egg case. Nautiluses have around 100 sucker-less limbs. All other cephalopods have 8 to 10 limbs that bear suckers and/or hooks.

DIVERSITY

More than 1,000 species of cephalopods exist in the world. Australia has a rich fauna, with the highest diversity of cuttlefishes and octopuses of any region. These molluscs occur in all seas and all habitats from polar regions, to tropical intertidal reefs and into the deep sea. Around 30 cephalopod species are common in Victorian coastal waters.

ECOLOGY

All cephalopods are carnivores. Lifestyles differ between the major groups. Open ocean squid are fast swimmers and tend to move in schools. The tiny bobtail and bottletail squids burrow in the sand during the day and emerge at night to feed. Pygmy squids use glue glands on their backs to glue under seagrass leaves. Cuttlefishes are expert at camouflage, matching both the patterns and textures of their surroundings. Octopuses are the most cryptic, using their elastic bodies to squeeze into nooks and crevices.

Southern Bobtail Squid, *Euprymna tasmanica* (Sepiolidae)

Southern Blueringed Octopus, *Hapalochlaena maculosa* (Octopodidae)

Maori Octopus, *Octopus maorum* (Octopodidae)

Giant Cuttlefish, *Sepia apama* (Sepiidae)

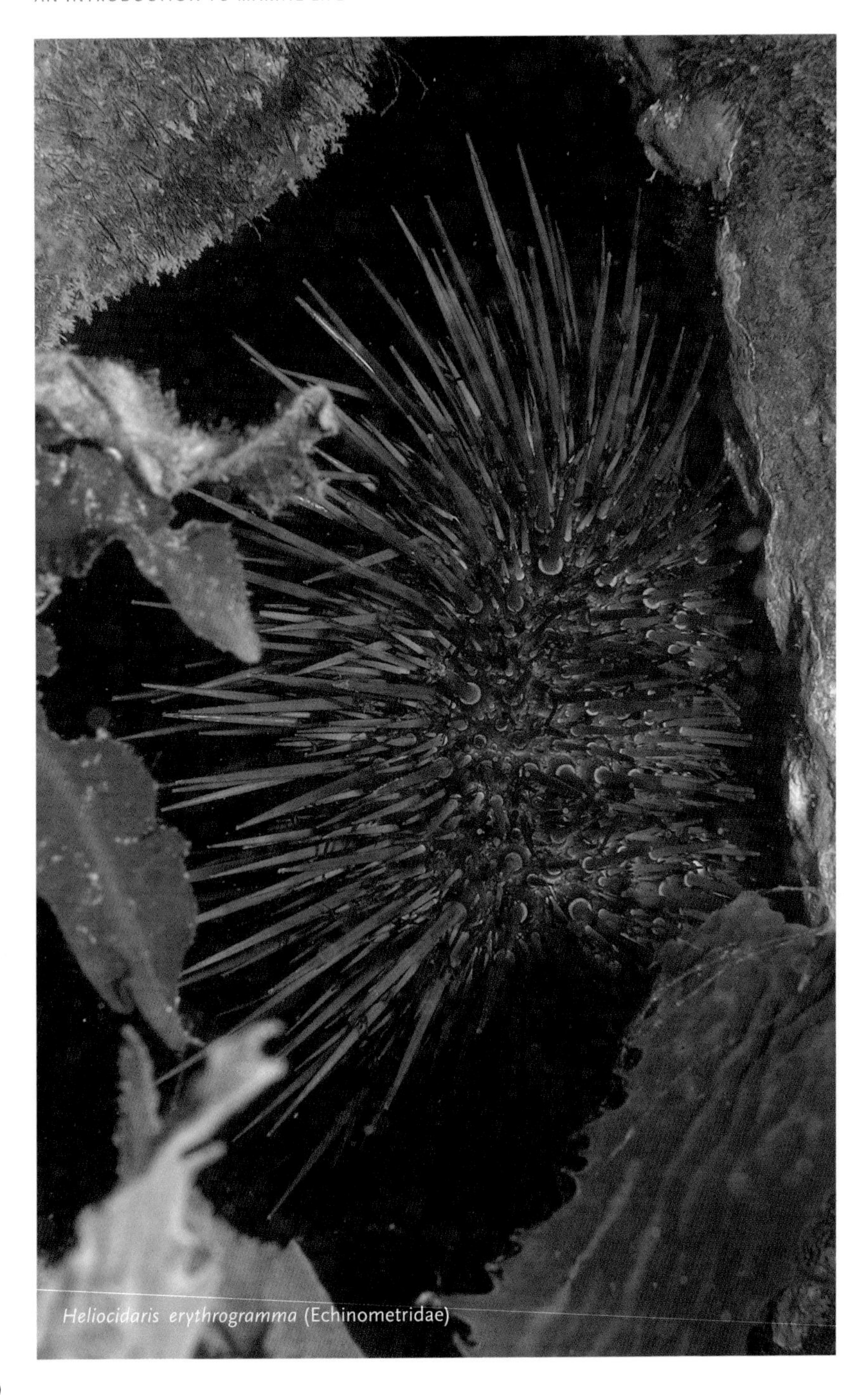

Heliocidaris erythrogramma (Echinometridae)

Sea stars, sea urchins and relatives (Phylum Echinodermata)

The Phylum Echinodermata includes sea stars, sea urchins, brittle stars, feather stars and sea cucumbers. These creatures may initially appear quite different, but they have inherited a unique set of characteristics: a star-shaped body plan (they are built around a five-armed or five-sided pattern), a skeleton of interlocking plates made of calcite crystals, and small fluid-filled pipes called tube feet. By watching a living animal in a rock pool or aquarium, you will usually be able to see the hundreds of tube feet, which most echinoderms use to move about. Echinoderms have no excretory organs and cannot regulate fluids and salts, so they only live in the sea (never in freshwater, or even estuaries). Most echinoderm species have separate sexes, but hermaphrodites are also known, especially among the brittle stars. Echinoderms can readily regenerate lost arms, and in sea stars and brittle stars asexual reproduction occurs in this way. Some holothuroids are also capable of asexual reproduction by division and regeneration. Sexual reproduction may involve release of eggs and sperm and external fertilisation, or fertilisation may be internal with brooding or more direct development of juveniles. The realisation that brooding and diverse reproductive strategies occur among southern Australian sea stars and sea cucumbers has led to the discovery of many new endemic species.

Tube feet of *Meridiastra gunnii* (Asterinidae)

Tosia neossia (Goniasteridae)

Cenolia trichoptera (Comasteridae)

Feather stars (Phylum Echinodermata, Class Crinoidea)

Crinoids are echinoderms with five or more long, feathery arms arising from five radial plates. Like all echinoderms, their skeleton is made up of calcite plates. The arms carry comb-like rows of pinnules that in turn have rows of tube-feet used to capture plantkonic prey. Some crinoids from deeper environments have long stalks which carry the feathery crown of arms. In tropical coral reefs crinoids are conspicuous and often seen by divers, but in temperate seas they are more cryptic, often remaining hidden in crevices during the day and extending their feathery arms to feed at night.

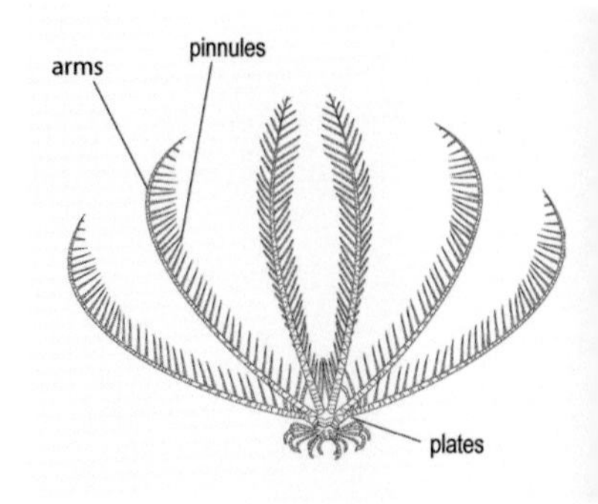

Generalised Feather star

IDENTIFICATION TIPS

Recognising a crinoid should be easy enough if the entire animal is on view, since there are no similar organisms in the sea. The long feathery pinnules on the arms readily separate crinoids from brittle stars. Sometimes divers see only a couple of crinoid arms extending from a cave; at first glance these may look like a branch of a hydroid colony but closer inspection will reveal the jointed calcite plates which enable crinoid arms to move and retract. Hydroid branches, though feathery in appearance, carry rows of polyps, have no plates or tube feet, and only move passively in the current.

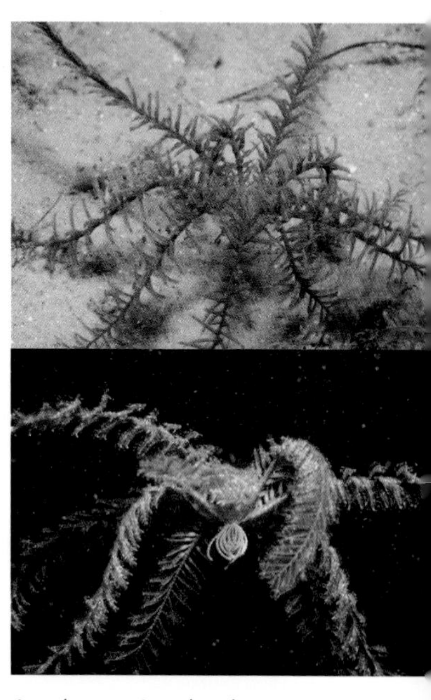

Antedon sp. (Antedonidae)

DIVERSITY

Crinoids are the least diverse of the echinoderm classes. More than 600 species are known worldwide and there are about 130 Australian species. Fewer than twenty species are presently known from southern Australia.

ECOLOGY

Crinoids are more likely to be found in deep water; in southern Australia they are mostly found on rocky reefs, often on bryozoans, sponges or gorgonians where they are ideally placed to capture plankton from passing water currents.

Uniophora granifera (Asteriidae)

Petricia vernicina (Asteropseidae)

Sea stars
(Phylum Echinodermata, Class Asteroidea)

Sea stars are perhaps the most widely known of all marine invertebrates. The radial star-like symmetry is plain to see and a hand lens will reveal intricately sculpted interlocking calcite plates. Turn over a sea star and the tube feet, also characteristic of all echinoderms, can be seen in rows under each arm. The mouth is under the disc, at the centre of the arms, and the anus is on the upper surface. Although most echinoderms have a five-armed body plan, some common sea stars in southern Australia have six, seven, eight or eleven arms.

IDENTIFICATION TIPS

The only sea stars that may be difficult to recognise are those in which the arms are reduced in size so that the animal has a cushion-like appearance (these species often have five to eight arms). However, the star-shaped symmetry and tube feet are evident when the sea star is turned over.

DIVERSITY

More than 1,800 species of sea star are described worldwide, and about 280 occur in Australian waters. About 50 species of sea star are known from southern Australia.

ECOLOGY

Most asteroids have specific substrate preferences. Species found on sandy bottoms are rarely encountered on rocky reefs, and vice versa. Some sea stars have no particular food preference and will eat whatever is available. Other species have preferences for algae, seagrass, encrusting invertebrates, molluscs and other echinoderms.

Coscinasterias muricata (Asteriidae)

Meridiastra sp. (Asterinidae)

Nectria sp. (Oreasteridae)

Tosia neossia (Goniasteridae)

Conocladus australis (Gorgonocephalidae)

Brittle stars

(Phylum Echinodermata, Class Ophiuroidea)

Ophiuroids are echinoderms which have a small, usually circular, central disc and long slender flexible arms. Most species have five arms, although in one group of brittle stars, the basket stars, the arms divide repeatedly to form a network of smaller branches. The mouth is at the centre of the disc, underneath. Unlike sea stars, brittle stars have no anus. The tube feet are reduced in brittle stars, which instead use their flexible arms for movement.

IDENTIFICATION TIPS

Brittle stars, with their long, flexible arms, are unmistakeable. However the arms are usually very fragile, and detached fragments of arm are sometimes mistaken for segmented polychaete worms. Close examination will reveal all kinds of difference between polychaetes and brittle star arms but the most obvious is that polychaetes do not crunch when squeezed with forceps!

DIVERSITY

Over 2,000 species of brittle star are known worldwide. Over 300 species occur in Australia, and about 100 species are known from the southern coast.

ECOLOGY

Brittle stars are mostly deposit feeders which consume surface detritus and scavenge other organic material, but a few species use their arms for filter feeding.

Ophionereis schayeri
(Ophionereididae)

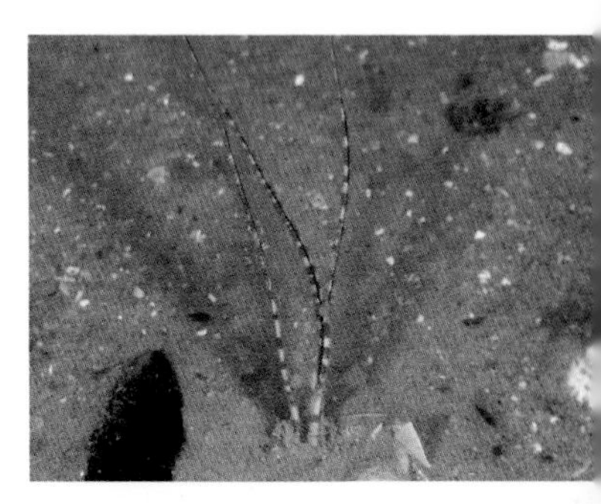

Unidentified brittle star
(Amphiuridae)

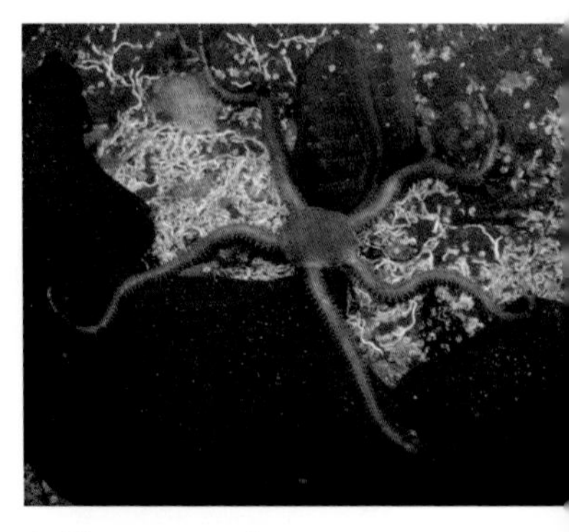

Ophiomyxa australis
(Ophiomyxidae)

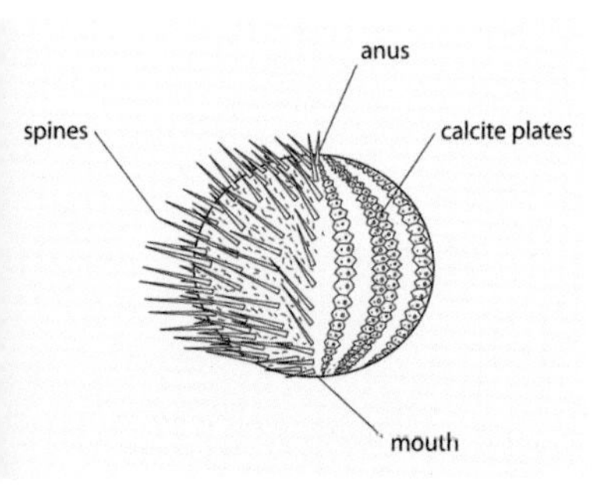

Generalised echinoid

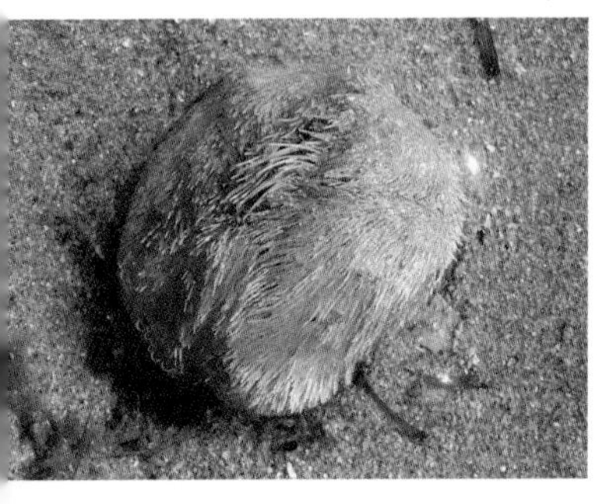

Echinocardium cordatum (Loveniidae)

Sea urchins
(Phylum Echinodermata, Class Echinoidea)

Sea urchins are echinoderms in which the interlocking calcite plates form a rigid shell (test). The spines in the different species range from short sharp bristles to long thin spines to smooth or serrated thick spines (as in the pencil and slate urchins). The five-rayed echinoderm structure can be seen by imagining a sea star with its arms folded above to meet, and with the plates interlocking to join the 'arms' together into a shell. Some plates have pores through which tube feet project, while others carry spines which are articulated on spherical tubercles; these two kinds of plates form separate alternating rows. Echinoids may either be 'regular' (with spherical test) or 'irregular' (often flattened). Echinoids have a central mouth underneath and an

Heliocidaris erythrogramma (Echinometridae)

anus on the upper surface. The chewing mouth has a set of five elongate vertical jaws held together in a structure known as Aristotle's lantern.

IDENTIFICATION TIPS

Sea urchins are so distinctive that they are unlikely to be mistaken for any other marine creature.

DIVERSITY

Approximately 800 species of sea urchin are known worldwide. At least 200 species occur in Australia, and 50 of these occur along the southern coast.

ECOLOGY

Most echinoids browse algae and seagrass and scavenge whatever animal food they can find. Regular echinoids mostly occur on rocky reefs and in seagrass, while irregular echinoids are burrowers adapted for moving through muddy and sandy sediments.

Goniocidaris tubaria (Cidaridae)

Centrostephanus rodgersii (Diadematidae)

Holopneustes inflatus (Temnopleuridae)

Sea cucumbers
(Phylum Echinodermata, Class Holothuroidea)

Holothurians are soft-bodied echinoderms which are slug- or worm-like. They are also known as sea cucumbers or (in tropical seas) as *trepang* or *bêche de mer*. Their similarity to other echinoderms is not immediately obvious yet when viewed end-on many can be seen to have a five-rayed symmetry, with tube feet arranged in five radial rows along the body. Some holothurians have irregularly arranged tube feet, or may lack them completely. The mouth is located at one end, and is surrounded by five or more branching oral tentacles (modified tube feet) and the anus is at the other end. The calcareous plates seen in other echinoderms are reduced to minute calcite ossicles, exhibiting a vast variety of form, embedded in the body wall.

IDENTIFICATION TIPS

Some holothurians are easily mistaken for several different phyla of unsegmented worms (see Quick Guide 3—Worm- and slug-like animals). Even experienced zoologists may need to dissect specimens to reveal internal structures to be sure of the correct phylum. Most unsegmented worms lack any sort of structures like the papillae and tube feet found on most holothurians. Among unsegmented worms, only Priapula (penis worms, p61) have an anus at one end and mouth at the other as in holothurians, but priapulans are rarely seen and, unlike holothurians, have a body divided into two distinct regions. Apodous holothurians (lacking tube feet) belonging to the order Apodida are not immediately recognisable as holothurians; these little red worm-like holothurians are quite common in shallow reefs and among algae. The most obvi-

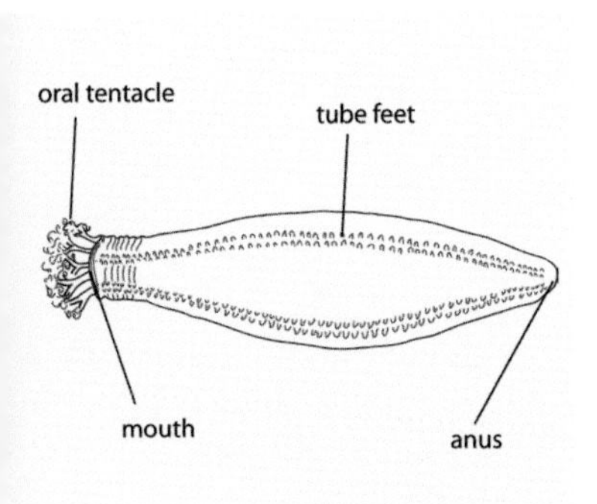

Generalised holothurian

Squamocnus aureoruber (Cucumariidae)

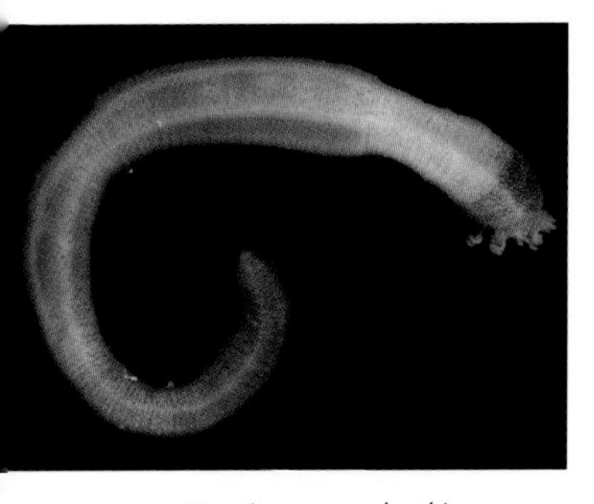

Taeniogyrus roebucki (Chiridotidae)

ous hint to the identity of Apodida is the spicules, which give them a 'sticky' feel when touched.

DIVERSITY

At least 1,400 species of holothuriaus are known worldwide. Over 200 species are known from Australia, and nearly half of these occur along the southern coast. Probably the most commonly seen shallow water holothuriaus in southern Australia is *Australostichopus mollis*, which is covered in irregular bumps, is usually brown-grey, and has tube feet along the ventral side only.

Australostichopus mollis (juvenile) (Stichopodidae)

ECOLOGY

Most holothuriaus are deposit feeders which consume sediment to extract organic material.

Australostichopus mollis (Stichopodidae)

Lace corals (Phylum Bryozoa)

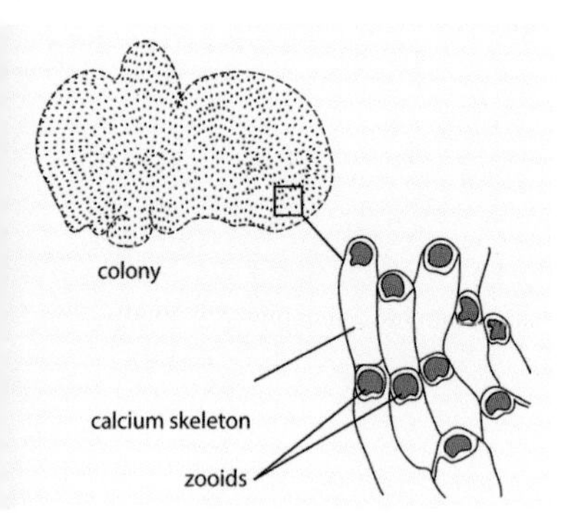

Generalised bryozoan

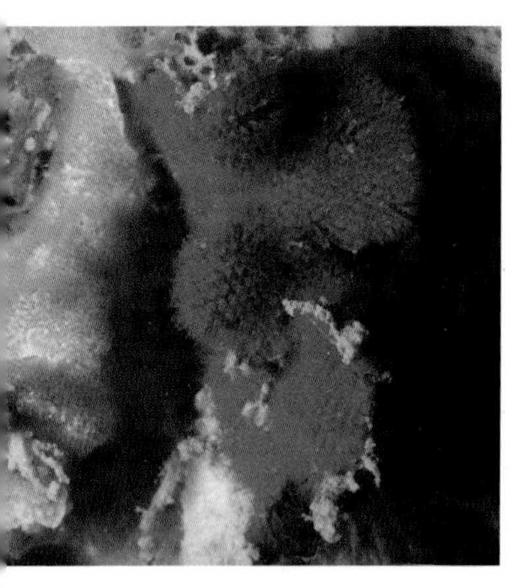

Mucropetraliella elleri (Petraliellidae)

Cellaria sp. (Cellariidae)

Most bryozoans are sessile colonial marine animals, and all feed using a circlet of tentacles to filter food from the water. There are a variety of forms, such as flat encrusting sheets, rigid branched upright structures, soft seaweed-like fronds, and distinctive lace-like clusters. Some forms resemble other marine organisms, giving the group the common name of moss animals or lace corals. The bryozoans are a notably colourful component of the marine fauna, with blues, oranges, reds, and vivid pink forms on display.

Colonial bryozoans are made up of tiny individual animals (zooids) which are fused together and anchored to a skeleton made of chitin or calcium carbonate. Each zooid is approximately one millimetre long and shaped like a box. The colony is made up of a few to millions of these tiny zooids; some colonies can reach metres in size. An anchoring structure (rhizoid) attaches the colony to the substrate, which is usually a hard rocky or shelly surface, but occasionally a soft surface such as seaweed. Some bryozoans are not attached to the substrate and are free to move slowly along the seabed.

IDENTIFICATION TIPS

The shape and colour of a bryozoan colony may be a quick way to identify a species. However, these features sometimes vary within a species, depending on the local habitat. Where the identification is uncertain, a microscope may be required to view the individual zooids.

DIVERSITY

There are more than 5,000 bryozoan species described worldwide. A few species live on the sand and a few in freshwater; the rest are marine. To date,

500 have been recorded from southern Australia.

ECOLOGY AND HABITAT

Most bryozoans are sessile, but smaller forms can move along the sea floor at a rate of up to one metre per hour. Bryozoans live from the intertidal zone to depths past 8,000 m. Large colonial bryozoans provide habitat for small marine animals such as polychaete worms, amphipods, copepods and other small crustaceans.

Zooids in the bryozoan colony feed using a circlet of tentacles (lophophore) which filters the water. Cilia on the lophophore capture food particles such as microalgae and bacteria, which are then propelled towards the mouth (at the top of the zooid) and into the U-shaped gut. When the lophophore is retracted, the top of the zooid may be covered by an operculum and is protected by spines in some species. There are no excretory organs, blood circulatory system or respiratory system. The zooids are small enough that gas exchange can take place across the body surface. Some zooids do not have a lophophore; they are structurally modified for other functions within the colony such as defence and cleaning, or as attachment surfaces and brood chambers.

Most bryozoans are hermaphrodites, either containing separate female and male zooids in the colony or containing both sexes within the one zooid. The larvae float in the plankton before settling onto a substrate and attaching using an adhesive substance. After developing into an adult, the zooid divides repeatedly to form the colony. The colony can grow quickly and thus can be a pest where it fouls thc hulls of ships or the wooden structures of piers.

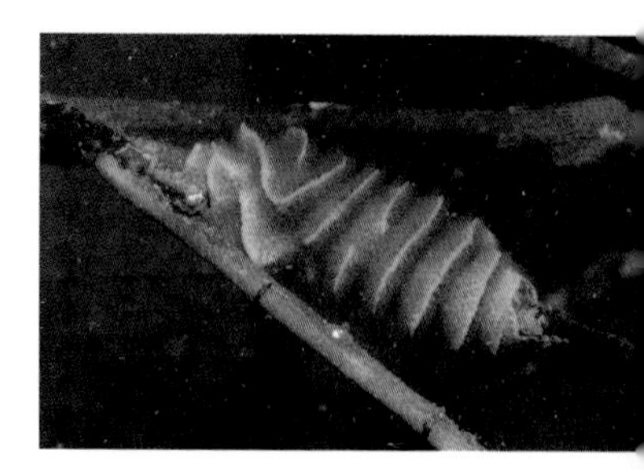

Celleporaria cristata (Lepraliellidae)

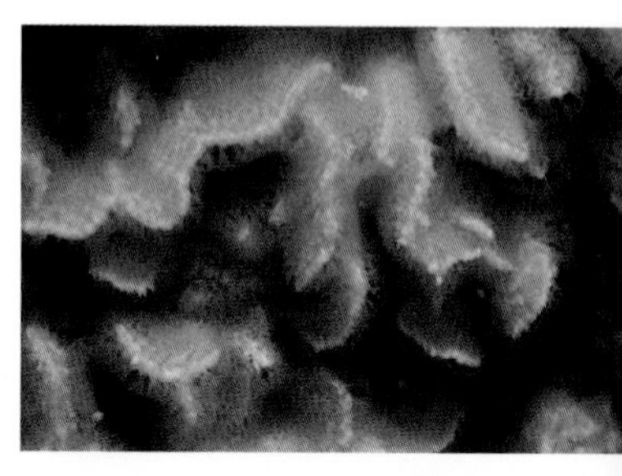

Celleporaria foliata (Lepraliellidae)

Orthoscuticella ventricosa (Catenicellidae)

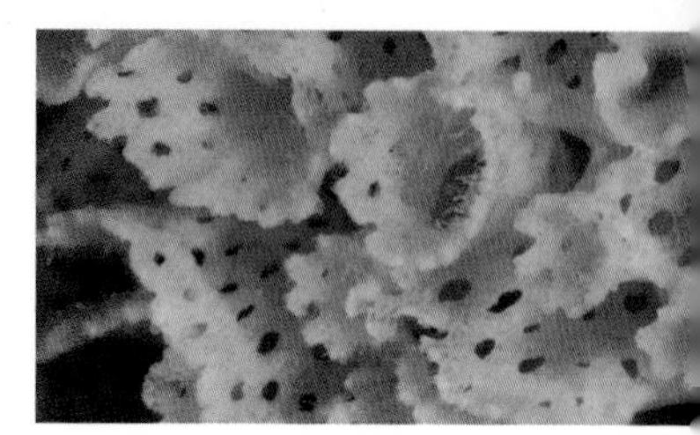

Triphyllozoon sp. (Phidoloporidae)

Lamp shells (Phylum Brachiopoda)

Brachiopods have a shell made of two valves. In nearly all species the valves are unequal and the upper valve has a hole through which a stalk attaches to the substrate. Inside the valves, brachiopods have a lacy membrane of ciliated filaments called a lophophore supported by an internal ring-like frame. The lophophore is a structure for filter feeding, but is also used for respiration, excretion and brooding young.

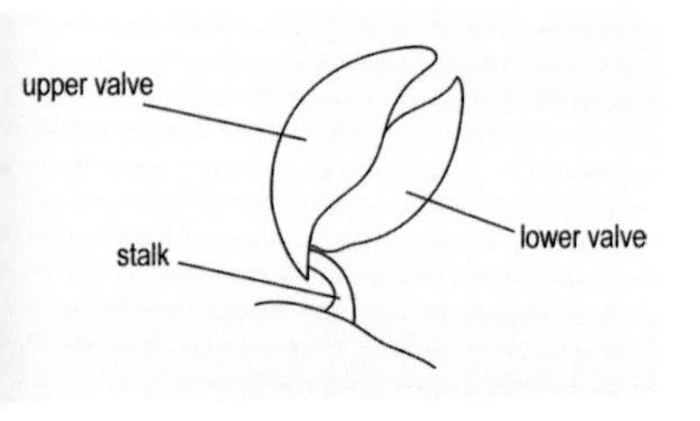

Generalised brachiopod

Magellania flavescens (Terebratellidae)

Magellania flavescens (Terebratellidae)

IDENTIFICATION TIPS

Brachiopods are superficially similar to clams and other bivalved molluscs, but they are quite unrelated. Their common name—lamp shells—alludes to the ancient Roman oil-burning lamps which they resemble. It has even been suggested that brachiopod shells inspired the design of oil lamps. The unequal, stalked valves of a brachiopod are distinctive, with the internal anatomy too very different from the animals they are most likely to be confused with: bivalve molluscs (Phylum Mollusca, Class Bivalvia, p86).

DIVERSITY

Brachiopods have a long and diverse history: over 30,000 fossil species are known, going back 600 million years. There are now only about 350 living species worldwide. However, they are quite widespread, and occasionally common, in Australian waters where almost forty species are recorded. The most common brachiopod in inshore waters of southern Australia is *Magellania flavescens*.

ECOLOGY

Brachiopods are filter feeders, and in Australia they are most common subtidally on rock or coarse shelly sediments, for example Bass Strait. Some brachiopod species are found on a wide range of sediments, but others are more specific and the stalk and associated structures may be adpated for attachment and life on different substrates.

Acorn worms, pterobranchs (Phylum Hemichordata)

Acorn worms (Class Enteropneusta) are marine worms with an anterior, smooth proboscis, and a longer, ridged or sculptured trunk region. Separating the regions is a distinctive collar. About 75 species are known worldwide; with at least eight species in Australia. They occur in many environments, but mostly among seagrass or mangroves.

Ptychodera flava (Ptychoderidae)

Minute filter-feeding organisms known as pterobranchs (Class Pterobranchia) are also classified in Phylum Hemichordata. These tiny creatures have tentacles covered with fine hair-like cilia, and also have a proboscis. Pterobranchs reach 5 mm in length and live in fine branching tubes. Few species are known worldwide. Only a single species, *Rhabdopleura normani*, is known from southern Australia (based only on an empty tube). See Further Information.

Arrow worms (Phylum Chaetognatha)

The side fins and long body give these 'arrow worms' their common name. They are grouped into a phylum of their own, but based on embryonic development are most closely related to echinoderms and chordates (tunicates, lancelets and vertebrates). They are fast predatory swimmers, with large eyes, strong jaws, and hooks surrounding their head. Most are less than 2 mm long and highly abundant in marine plankton; and 70 known species worldwide.

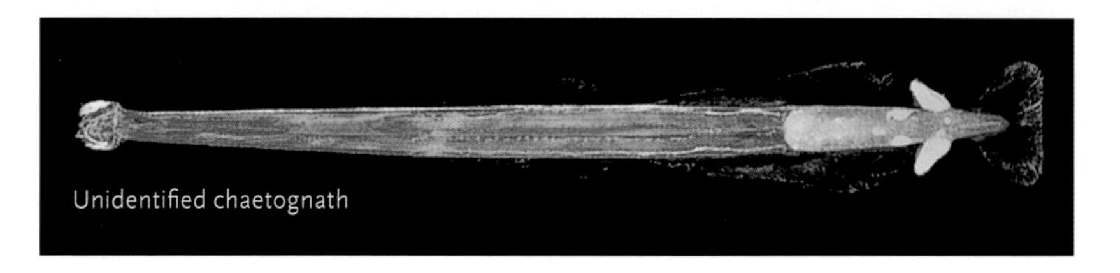

Unidentified chaetognath

Ornate Cowfish, *Aracana ornata* (Aracanidae)

Mosaic Leatherjacket, *Eubalichthys mosaicus* (Monacanthidae)

Vertebrates and relatives (Phylum Chordata)

The Phylum Chordata includes the vertebrates (familiar animals like fish, lizards, dogs, whales and of course humans). However this phylum also includes several less familiar forms, such as the small, fish-like lancelets, sea squirts, and planktonic jelly-like salps. It is not immediately obvious why these disparate organisms should be classified together, nevertheless careful study, especially of larvae, shows that all possess a nerve chord along their back above a skeletal rod (notochord), and gill slits. The adult and larval forms can look very different, and some of the typical chordate features are lost in adults of some groups.

Australian Pelican, *Pelecanus conspicillatus* (Pelecanidae)

Lancelets (Phylum Chordata, Subphylum Cephalochordata)

Lancelets (amphioxus) live in shallow seas where they burrow into the sand and poke their head out to catch food particles. Their bodies are elongate, with pointed ends, and they are usually about 50 mm long. Along with tunicates and vertebrates, the lancelets are chordates, possessing a nerve chord along their back above a skeletal rod (notochord), and gill slits. Unlike the vertebrates however, they lack a true backbone. There are only two species known from southern Australia and 36 species known worldwide.

Lancelet, *Branchiostoma lanceolatum* (Branchiostomidae)

Botrylloides magnicoecus (Styelidae)

Clavellina sp. (Clavellinidae)

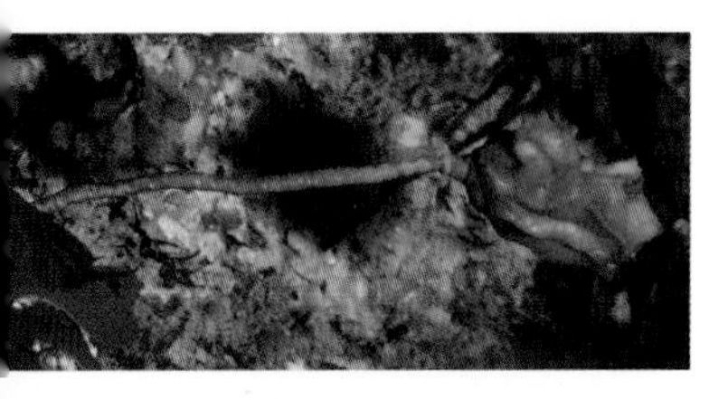

Pyura sp. (Pyuridae)

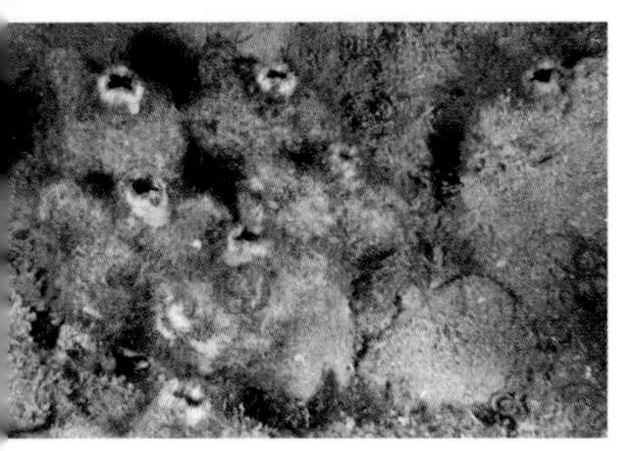

Pyura stolonifera (Pyuridae)

Sea squirts

(Phylum Chordata, Class Ascidiacea)

The sea squirts (ascidians) are named for their ability to forcefully expel water from a large pore. This may be particularly obvious in those sea squirts exposed on the rocky shore during low tide. The larvae are free-swimming and do not feed. They only live for a few hours, swimming about looking for a place to settle. There they change into the adult form, growing a sack-like head and a stalk. The sea squirt takes in water through a tube (inhalant siphon), filters it for food and then expels it through another tube (exhalent siphon). In some species, the individual animals are very small and live together in a colony, with different animals specialised for different functions.

IDENTIFICATION TIPS

Distinguishing ascidians from other encrusting marine invertebrate growths is easy enough if only the distinctive pair of siphons can be recognised (see Quick Guide 2). To identify sea squirts, investigate whether the animal is solitary, in a loose aggregation, or in a colony of tiny animals (individuals in the colony are only 1 mm in size). Species differ in the length of their stalk, with some having an obvious stalk and flower-shaped head (sea tulips) and others having no visible stalk at all. The shape of the sea squirt varies from simple sack, to long lobes or mats in colonial forms. There are variations in ridges, spots, furrows, processes and colour.

DIVERSITY AND ECOLOGY

Sea squirts play an important ecological role, filtering tiny food particles such as bacteria and phytoplankton from the water. Most sea squirts live on rocky reefs, in soft sediments on the sea floor, or attached to seagrasses. The outside of the sea squirt may be home to algae, bryozoans and sponges. There are 2,000 species worldwide and 300 species known from southern Australia. Sea squirts are hermaphrodites, but the male and female reproductive organs mature at different times so self-fertilisation is unusual. Fertilisation is typically external in solitary ascidians, but internal (with subsequent brooding of larvae) in colonial species. Colonial species increase the size of the colony by budding new individuals.

Scyozoa cerebriformis (Holozoidae)

Salps

(Phylum Chordata, Class Thaliacea)

The semi-transparent salps live in the open ocean, drifting in ocean currents across vast distances. They may group together to form colonies and can form large swarms when conditions are favourable. Like other tunicates, they filter food particles from the water. They are eaten by seabirds, cnidarians (jellyfish), molluscs and fish such as tuna. There are 70 species known worldwide and 20 from southern Australia. You might find a salp washed up on the shore or sparkling in the water at night: *Pyrosoma atlanticum* occurs in Southern Australia, is bioluminescent and is sometimes visible in shallow waters.

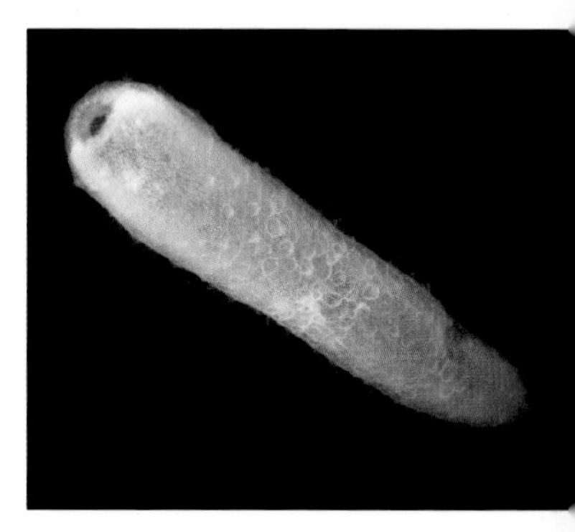

Pyrosoma atlanticum (Pyrosomatidae)

Pegea sp. (Salpidae)

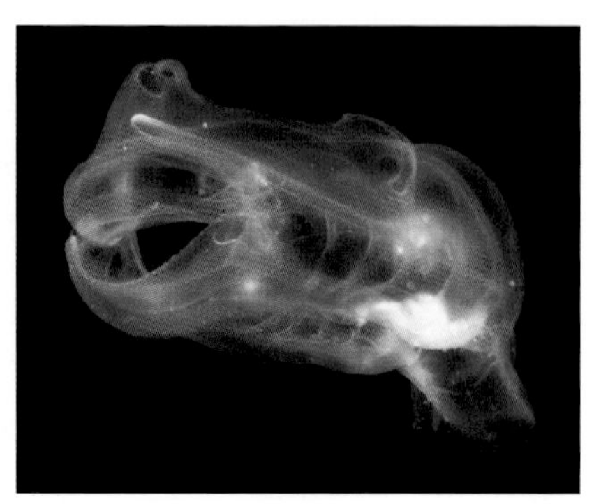

Pegea confoederata (Salpidae)

Pouched Lamprey, *Geotria australis* (Pteromyzontidae)

Lampreys (Phylum Chordata, Class Cephalaspidomorphi)

Lampreys are slimy eel-like fishes that lack jaws, feeding instead with a rasping tongue. They are thought to have descended from armoured fishes that were around before the dinosaurs. Lampreys have seven small gill slits and a single nostril on top of the head. They have a cartilage skeleton, lack scales and only have dorsal and anal fins. They can reach up to 600 mm long. There are 31 lamprey species in the world with two marine species in southern Australia. Pouched Lampreys (*Geotria australis*) have a mouth fringed by fleshy papillae and spend most of their life in freshwater. Only immature adults occur in the sea. Shorthead Lampreys (*Mordacia mordax*) lack the fringe of papillae around the mouth and can occur in the sea as immature adults.

Longfin Hagfish, *Eptatretus longipinnis* (Myxinidae)

Hagfish (Phylum Chordata, Class Pteraspidomorphi)

Hagfish have many similarities with lampreys—slimy, eel-like, lacking jaws and with a cartilaginous skeleton—they were once classified together. They are now known to have separate origins. Two species of hagfish have been recorded in southern Australia with the Longfin Hagfish (*Eptatretus longipinnis*) being the one most likely to be seen in Victoria and South Australia. It has two pairs of barbels (feelers) at the tip of the snout, no dorsal fin, and six small gill slits on each side. The eyes are not obvious, being covered under a layer of skin. The Longfin Hagfish feeds by attaching to fish (or rock lobsters trapped in craypots). Unlike lampreys, hagfish do not enter freshwater streams.

Chimaeras (Phylum Chordata, Order Chimaeriformes)

Chimaeras are a group of primitive fishes closely related to the sharks and rays. These fishes share a skeleton that is all or mostly made of cartilage. Chimaeras tend to have smooth skin with an obvious network of lateral lines and sensory canals over the head. They also possess long tails, large eyes and fused beak-like jaws with grinding plates for crushing hard-shelled prey.

There are three different families of chimaeras, all represented in southern Australian waters: Ghost sharks (family Chimaeridae), Elephantfish (family Callorhinchidae) and Longnose chimaeras (Rhinochimaeridae). Ghost sharks are deepwater animals with a rounded head or short snout, and a thin tail that ends in a point or filament. Elephantfish can occur in coastal waters. They have a hoe-like hooked snout and a shark-like tail. The Longnose chimaeras are easily recognised by a long pointed snout and a thin tail that ends in a point or filament.

All chimaeras lay their young in special horny egg cases that tend to be teardrop-shaped with ribbed side flanges. The young animal is left to grow within this egg case for up to a year before hatching to lead an independent life.

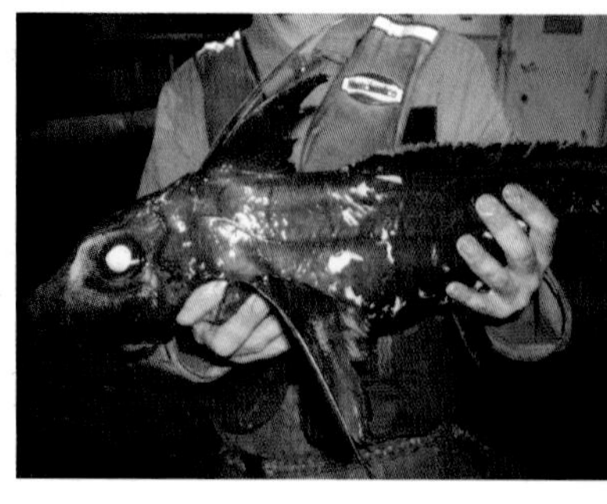

Black Chimaera, *Hydrolagus* sp. (Chimaeridae)

Chimaera hatchling (*Hydrolagus* sp.) and egg case from deep water in the Tasman Sea (Chimaeridae)

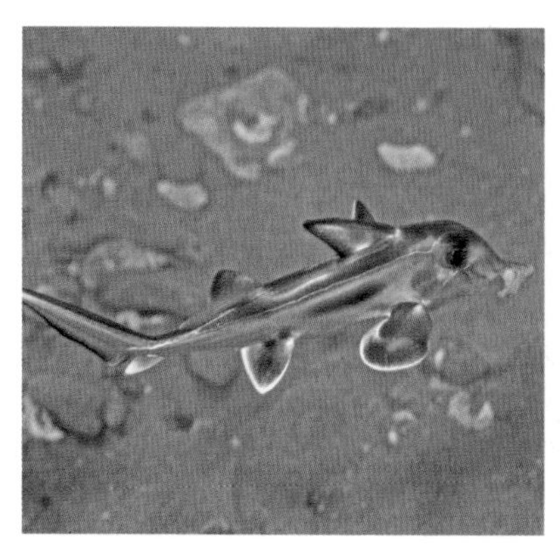

Elephantshark, *Callorhynchus milii* (Callorhynchidae)

Longnose Chimaera hatchling, *Rhinochimaera* sp. (Rhinochimaeridae)

Sharks and Rays (Phylum Chordata, Class Chondrichthyes)

Sharks and rays are different from bony fishes. The sharks and rays have a skeleton made of cartilage while the bony fishes have at least some calcified bones. Their teeth are highly modified versions of the denticles that give their skin a rough sandpaper-like texture. The teeth of sharks and rays fossilize well, so we know that they have a much longer evolutionary history than bony fishes. Some sharks, such as the White Shark, seem to have changed little over hundreds of millions of years.

IDENTIFICATION TIPS

It is easy to distinguish bony fishes from cartilaginous sharks and rays—the mouth of most sharks and rays is located under the head, away from the tip of the head, while in bony fishes the mouth is typically right at the front. However, many sharks and stingrays are distinctive enough that they are not easily mistaken. Smaller sharks such as catsharks are less familiar, but their sandpaper-like skin is very different from the scaly (or smooth) skin of most bony fishes (leatherjacket fishes being one exception).

Spotted Stingaree, *Urolophus gigas* (Urolophidae)

DIVERSITY

More than 100 species of sharks and rays have been recorded from southern Australian waters. The Gummy Shark, *Mustelus antarcticus*, is the species most familiar to Australians, since it is widely sold in fish shops as 'flake'—it only occurs in southern Australia, and this is also true of many other species of our sharks and rays. Rays are more often seen than sharks by snorkellers and divers, and large rays of the genus *Dasyatis* sometimes frequent piers where they may be visible as large dark moving shapes.

Australian Angelshark, *Squatina australis* (Squatinidae)

ECOLOGY

Only a few of the largest shark species eat large mammals and may be a danger to humans. Most are fish eaters. Skates and rays mostly consume invertebrates, especially molluscs which they can crush with their small vice-like jaws. Reproduction in sharks and rays involves copulation and internal fertilisation; subsequently they may lay eggs or give birth to live young (as does the Gummy Shark, for example). Some species are threatened by humans, for example the fish eating Greynurse Shark (*Carcharias taurus*) was once found in Port Phillip Bay but no longer occurs there. Despite having a wide distribution globally, the Greynurse Shark is critically endangered in eastern Australia.

White Shark, *Carcharadon carcharias* (Lamnidae)

Gummy Shark, *Mustelus antarcticus* (Triakidae)

Scalyfin, *Parma victoriae* (Pomacentridae)

Bony fishes
(Phylum Chordata, Class Pisces)

Like sharks and rays, bony fishes are vertebrates, so also have a backbone. However, while sharks and rays have a skeleton made of cartilage, the bony fishes have at least some calcified bones. A relative handful of bony fish species are well known by anglers, or from seafood restaurants and markets. However our waters contain hundreds of rarely seen fish species that remain hidden in rocky

Slender-spined Porcupine Fish, *Diodon nicthemerus* (Diodontidae)

reefs, seaweeds or buried in sandy bottoms. Many have fantastic camouflage while other fishes, such as Eastern Australian Salmon (*Arripis trutta*) and Jack Mackerel (*Trachurus declivis*) confuse potential predators by swimming in vast schools.

IDENTIFICATION TIPS

Fishes are probably familiar enough that they are easily recognised. They are only likely to be confused with sharks and rays, but the mouth of bony fish is typically positioned at the front of the head while in most sharks and rays the mouth is located under the head, further back. Some smaller eels or other fishes (such as snake blennies and eel blennies) can be mistaken for worms but are easily recognised by their obvious eyes, mouth, and one or more gill openings. For highly camouflaged fish such as the Tasselled Anglerfish, fins, eyes and mouth are the signs that you are looking at a fish not a rock covered in weed.

DIVERSITY

More than 1,100 species of fishes have been recorded from southern Australia. The majority are endemic to southern Australia, such as seadragons, Warty Prowfish (*Aetapcus maculatus*) and Velvetfish (*Aploactisoma milesii*).

ECOLOGY

Fishes show a huge variety in form and lifestyle. Tiny clingfishes have modified suckers that act as suction caps to grip rocks while waves surge around them. Worm eels dive into the sand to avoid predators. Recreational and commercial fishing pressure can affect fish populations, particularly close to the major cities. Reproduction in most bony fishes involves release of eggs and sperm to fertilise externally, but some species have evolved various kinds of care, some even brooding their eggs in their mouth.

Sand Flathead, *Platycephalus bassensis* (Platycephalidae)

Ornate Cowfish, *Aracana ornata* (Ostraciidae)

Sculptured Seamoth, *Pegasus lancifer* (Pegasidae)

Yellow-bellied Sea Snake, *Pelamis platurus* (Hydrophiidae)

Reptiles

(Phylum Chordata, Class Reptilia)

Besides the fishes, many other animals with backbones have a marine existence. Reptiles are one such group, but as they mostly rely on the sun to maintain body temperature there are few that reach the cool waters of southern Australia. However, turtles and sea snakes are occasional visitors from tropical waters.

IDENTIFICATION TIPS

At sea, marine reptiles are difficult to see, let alone identify. All are air breathers, so must return to the surface regularly to breathe. When swimming at the surface, turtles raise their head slowly in a distinctive awkward way. Sea snakes are rare visitors to our region, typically only found once washed ashore. Sea snakes can be distinguished from eels by the absence of gill openings and the coarse obvious scales. Sea snakes have paired fangs on the roof of the mouth while eels do not.

Leatherback Turtle, *Dermochelys coriacea* (Dermochelyidae)

DIVERSITY

Of the seven sea turtles in the world, only one species, the Leatherback Turtle (*Dermochelys coriacea*), regularly visits southern Australia. Of the 33 sea snakes found in Australia, the Yellow-bellied Sea Snake (*Pelamis platurus*) is a rare visitor to Victorian waters, in pulses of warm water coming down the east coast from the tropics.

ECOLOGY

Leatherback turtles mostly eat jellyfish at the surface and in deeper water. The Yellow-bellied Sea Snake eats small fish; the same venom that immobilises fish is potentially lethal to humans, so treat them with care. First aid as for other snake bites: immobilise and apply pressure to the limb with a tight bandage, seek medical attention urgently and bring the snake for venom identification if possible. Apply emergency resuscitation if necessary. Marine turtles lay eggs in tropical sand dunes at night. The Yellow-bellied Sea Snake gives birth to live young at sea. However, no marine reptiles breed in southern waters.

Birds (Phylum Chordata, Class Aves)

Numerous birds depend on the sea for some part of their existence. Some, like gulls and cormorants, are mostly confined to nearshore coastal waters. Others, like albatross and petrels, are birds of the Southern Ocean that breed in the Antarctic or Subantarctic and are rare in nearshore waters of southern Australia. One bird is flightless with a fully aquatic lifestyle. This is the Little Penguin (*Eudyptula minor*), the smallest penguin in the world. Although some seabirds may remain at sea for many weeks, all need land for breeding and for moulting and growing new feathers.

Pacific Gull, *Larus pacificus* (Laridae)

IDENTIFICATION TIPS

Birds are easily recognised—nothing else has feathers! Distinguishing one seabird from another is more difficult and will require one of the handbooks listed in Further Information below. Penguins sit very low in the water compared with other seabirds such as cormorants and gulls—they are often heard before they are seen. From a distance Little Penguins may look like Hoary-headed Grebes (which are also common in coastal waters). However, grebes can fly.

DIVERSITY

Around 700 bird species are known from Australia, and about 80 of these are considered seabirds. They regularly or occasionally visit southern waters. The Little Penguin is the only one of the 17 penguin species in the world that breeds in southern Australian waters. Only one albatross, the Shy Albatross, breeds in southern Australia (on islands around Tasmania). The Black-browed Albatross is more commonly seen in our waters, particularly in winter. Other common seabirds of coastal waters include two gulls (Silver and Pacific), four cormorants (Pied, Little Pied, Black-faced and Little Black), the Australasian Gannet and the Australian Pelican.

ECOLOGY

The principal prey of most seabirds are small fish such as pilchards (*Sardinops neopilchardus*, see Quick Guide 6, but squid and crustaceans are also taken in large numbers. Many seabirds will scavenge any floating carcass or other food at sea. Penguins lay eggs in rock crevices or in burrows in sand dunes. The Little Penguin returns from sea at dusk, often in large numbers, as occurs at the 'Penguin Parade' at Phillip Island—a popular tourist destination in Victoria. Reproduction in most other seabirds follows a similar pattern, with many relying on islands, inaccessible cliffs or even artificial marine structures to provide safe nesting sites.

Little Penguin, *Eudyptula minor* (Spheniscidae)

Shy Albatross, *Thalassarche cauta* (Diomedeidae)

Australian Fur Seals, *Arctocephalus pusillus* (Phocidae)

Mammals

(Phylum Chordata, Class Mammalia)

There are two main types of marine mammals in southern Australia: the pinnipeds (Suborder Pinnipedia, seals and sea lions) and the cetaceans (Order Cetacea, dolphins and whales). Although both are mammals—they have mammary glands and a placenta—cetaceans and pinnipeds are not closely related to each other. Pinnipeds belong to the mammalian Order Carnivora and so they are related to cats and dogs. The closest relatives of cetaceans are ungulates—pigs, deer, hippopotamuses and their relatives.

IDENTIFICATION TIPS

Recognising between seals, dolphins and whales, even at sea, is not difficult as long as they are close. Seals are often visible when they roll at the surface and raise their side or tail flippers in the air. Dolphins rise regularly to breathe, in a smooth curved action. Distinguishing species is another matter. Two seals are common in southern Australia: the Australian Fur Seal (*Arctocephalus pusillus*) and the New Zealand Fur Seal (*Arctocephalus forsteri*). Telling them apart is difficult even when they are seen on the shore. Two dolphins are also common in our waters: Bottlenose Dolphin (*Tursiops truncatus*) and Common Dolphin (*Delphinus delphis*). The Common Dolphin has more complex colour patterns

Australian Fur Seal, *Arctocephalus pusillus* (Phocidae)

Bottlenose Dolphin, *Tursiops truncatus* (Delphinidae)

(long sweeping bands of grey, black and white along the sides) so it can be recognised up close. Whales typically blow spray into the air, the shape of the spray and the shape of the back and fin can help in identifying the species. The two common large whales are the Humpback Whale (*Megaptera novaeangliae*) which has a small dorsal fin, and the Southern Right Whale (*Eubalaena australis*) which has none.

DIVERSITY

Ten species of seals and sea lions have been recorded from southern Australian waters, but most breed on subantarctic islands or Antarctica and rarely reach mainland Australia. Australian Fur Seals breed on Victorian offshore islands and haul-outs on channel markers in Port Phillip Bay. Elsewhere in southern Australia, and in New Zealand, New Zealand Fur Seals are more common. Elephant Seals (*Mirounga leonina*) once bred along our coastline and young males are starting to return. More than 40 species of dolphins, porpoises and whales feed, breed or travel through Victorian waters. The most commonly seen are the resident Bottlenose Dolphin and Common Dolphin, both of which may form vast aggregations moving along the coast. Their relatives, the toothed whales, include Sperm Whales (*Physeter catodon*) and

Pilot Whales that typically occur over deeper water off places such as Eden and Portland. Baleen whales all share long hairy plates in their mouth that they use when feeding. They include the Southern Right Whale, Pygmy Blue Whales (*Balaenoptera musculus brevicauda*) and Humpback Whales.

ECOLOGY

Fur seals mainly eat squid, octopus and fish. Dolphins eat squid, other cephalopods and fish; Bottlenose Dolphins even feed on lobsters. Toothed whales take larger squid and fish while the baleen whales use their baleen plates to sieve tiny shrimp-like krill from the water. Like seabirds, seals take to dry land to raise their young and to moult, when they grow a new waterproof fur coat. Whales and dolphins, of course, are totally marine and have evolved to carry out their whole life-cycle in the sea, from mating and birth to feeding their young milk; although they may seek sheltered coasts during breeding. Seals and sea lions need to come ashore to give birth and care for their young; such breeding colonies are typically on islands. Southern Right Whales come to Warrnambool every year to breed, while Pygmy Blue Whales feed on krill at upwellings off nearby Portland. Humpback Whales migrate along the east and west coasts of Australia to breed each winter.

Southern Right Whale, *Eubalaena australis* (Balaenidae)

ACKNOWLEDGEMENTS

For a book such as this that attempts a very broad coverage of marine life, acknowledgements are more numerous, and more important, than is perhaps normal. Firstly, we must thank our sponsors. Funding for the Museum Victoria Guides to Marine Life, of which this is the first, is gratefully acknowledged from the Australian Government through a grant from its Natural Heritage Trust. The grant was facilitated by the Port Phillip and Westernport Catchment Management Authority. Publication of the series is funded in part by the Victorian Coastal Council and Parks Victoria. Our publisher, Patty Brown, was tolerant, stern, cheerful and professional; usually all at once. Our thanks also to Marija Bacic, Image Managment and Copyright Assistant at Museum Victoria and Gary Presland for completing the indexes. This book has benefited from feedback from workshops and laboratory sessions involving members of the Marine Research Group of the Field Naturalists Club of Victoria, ReefWatch Victoria, and third year classes of the Department of Zoology of The University of Melbourne. Very many of our friends and colleagues at Museum Victoria and many other institutions have corrected errors, made suggestions and each did their best to ensure that their own field of expertise was treated accurately.

Photographs and illustrations

We are grateful to the many photographers and illustrators who made their work available for publication.

John Ahern 15(9), 82(T) Kelvin Aitken 25(12), 113C Leon Altoff 11(16), 59TR Kathie Atkinson 20(3) Jon Augier 20(4) Peter Batson 110B, 105B, 69B Phil Bock 12(1)inset Ian Bolch 16(6) William Boyle 28, 99B, 97B Dean Chamberlain 121 John Chuk 54B, 54C, 15(15), 56B, 85BR, 15(13), 54T CSIRO 26(1) and (2) Kyatt Dixon 34, 35T,35CL, 10(4), 20(1), 35BR Roger Fenwick 25(13), 44, 48B, 17(14), 108UC, 9, 96 Julian Finn 91T, 25(7), 106T, 11(18), 95T, 95B, 101B, 25(11), 122, 86C, 49CR, 115C, 91B, 12(3), 49BL, 10(8), 103T Karen Gowlett-Holmes 27(8), 49BR, 21(9), 82B, 83T, 60T, 15(14), 60B, 23(7) T. Joan Hales 85BL, 82CL, 15(11), 117, 17(15), 98C, 21(10), 18(7), 14(2), 79BR, 14(5), 57B Gustaaf Hallegraeff 30B Simon Heislers 27(5) Ian Kirwan 100C, Rebecca Koss 35BL Damon Kowarsky 57T Rudie Kuiter 110T, 111BL, 113B, 21(11), 85CR, 21(6), 16(5), 22(1), 47B, 88T, 16(1), 22(3), 50B, 104B Gary Lewis 118 Lindy Lumsden 123 Richard Marchant 64B Michael Marmach 15(12), 100B, 75T, 74B, 72B, 27(6), 59TL, 20(2), 25(9), 23(8), 68inset, 25(10), 21(8), 27(4), 79TL, 77T, 77LC, 21(7), 18(6), 79BL, 78B, 75B, 70T, 72T, 69T, 68, 71B, 77B, 66C, 59BR, 65, 87T, 74C, 71T Peter Marmach 23(12), 99T, 73UC, 14(8) Loisette Marsh 105T Brad Maryan 116B Anna McCallum 74T, 78T, 104T, 102T, 98T, 93T, 46T, 73T, 50T, 76T, 100T, 66T Jasper Montana 112Florence V. Murray 19(19) Andrew Newton 14(1) Mark Norman 19(10), 18(9), 18(8), 48C, 119, 111BR, 111T, 111C, 25(8), 51T, 16(3), 84B, 20(5), 81, 108B, 11(17), 108LC,

87B06B, 95T, 95LC, 97T, 14(7), 10(6), 40C, 87CR, 43BR, 86B, 76B, 18(5), 2, 49T, 94T, 39T, 114B, 3, 27(9), 94inset, 26(3), 39BL, 37B, 95UC, 101T, 10(5), 40B, 41B, 19(16), 17(13), 109C, 17(11), 109B, 45, 73B, 98B, 87CL, 18(3), 5, 42T, 17(9), 120, 14(3), 58B, 52B, 24(2), 10(7), 33 David Paul 17(8), 52T, 16(2), 22(1), 50C Gary Poore 40T Peter Robertson 116T Chris Rowley 4, 15(10) David Staples 84T, 67T, 83B, 13(4), 75CL, 15(16), 61T, 31, 66B, 67B Erling Svensen 61B Rudolf Svensen 107B Anna Syme 70B Platon Vafiadis 56T, 24(3), 27(7), 86T Jeanette Watson 12(1) Kim Westerkov 17(12) Robin Wilson 13(6), 13INSET, 10(1), 29L, 38T, 79TR, 10(2), 32B, 107T, 18(2), 18(4), 38B, 19(17), 18(1), 19(18), 19(13), 19(14), 19(12), 19(15), 84C, 10(3), 32T, 16(4), 47LC, 85CL, 29R, 19(11), 75CR, 13(7), 78C, 91T

PHOTO CREDIT LEGEND	
Top	T
Top Left	TL
Top Right	TR
Centre	C
Centre Left	CL
Centre Right	CR
Bottom	B
Upper Centre	UC
Lower Centre	LC
Bottom Right	BR
Bottom Left	BL

ABOUT THE AUTHORS

Robin Wilson is Senior Curator of Marine Invertebrates at Museum Victoria. He has a life-long fascination with marine life, especially creatures that are less well-known. In the 1980s he studied polychaete worms–one of the most common of all marine animals–and has since concentrated on understanding the diversity and taxonomy of these ecologically important organisms. Robin is also interested in the evolution and biogeography of southern Australian marine life, and in building partnerships with amateur naturalists and other community groups.

Mark Norman is Head of Sciences at Museum Victoria. His main research interests are the behaviour, biology and evolution of cephalopods—the octopuses, squids, cuttlefishes and nautiluses. Mark has completed field surveys for these animals throughout Australia and the world. His other interests include fishes, other southern Australian marine life, Antarctica and science communication. Mark has been a film maker and the author of numerous books and articles. He is an award-winning children's book author and recently published books include *The Koalas: The Real Story* and *The Great Barrier Reef Book: Solar powered.*

Anna Syme recently finished her PhD at Museum Victoria, studying small crustaceans known as ostracods. She currently works at the University of California, Santa Barbara in the USA and investigates evolutionary biology using ostracods as a study animal. Anna is interested in marine biodiversity and molecular biology.

CLASSIFICATION OF MARINE LIFE

The classification presented below includes the main groups in the body of the text, as well as many marine life forms not mentioned in the text (because they are microscopic, or extinct, or not known from Australian seas). Biology texts differ. The Tree of Life web pages (http://tolweb.org) and other reference sources quoted in Further Information, can be consulted for updates and for details of land- and freshwater-dwelling organisms not covered here.

Phylum Cyanobacteria (blue-green algae; see text p32)

Phylum Bacillariophyta (diatoms; see Further Information—*microbes, protists* p132)

Phylum Phaeophyta (brown algae; see text p38)

Phylum Chrysophyta (rhaphidophytes; see Further Information—*microbes, protists* p132)

Phylum Bacillariophyta (dinoflagellates; see text p30)

Phylum Rhodophyta (red algae; see text p36)

Phylum Chlorophyta (green algae; see text p34)

Phylum Anthophyta (flowering plants; see text p40)

Phylum Haptophyta (prymnesiophytes; see Further Information—*microbes, protists* p132)

Phylum Actinopoda (radiolarians; see Further Information—*microbes, protists* p132)

Phylum Foraminifera (=Phylum Granuloreticulosa; forams; see text p31)

Phylum Ascomycota (fungi, including lichen-forming fungi; see text p29)

Phylum Porifera (sponges; see text p42)

Phylum Cnidaria (corals, anemones, jellyfish; see text p45)

- Class Hydrozoa (hydroids and relatives; see text p46)
- Class Anthozoa (sea anemones, corals, sea pens; see text p48)
- Class Scyphozoa (jellyfish; see text p50)
- Class Cubozoa (box jellies; see text p50)

Phylum Ctenophora (comb-jellies; see text p53)

Phylum Platyhelminthes (flatworms; see text p55)

Phylum Rotifera (rotifers; see Further Information—*planktonic and interstitial forms* p133)

Phylum Gastrotricha (gastrotrichs; see Further Information—*planktonic and interstitial forms* p133)

Phylum Kinorhyncha (kinorhynchs; see Further Information—*planktonic and interstitial forms* p133)

Phylum Nemertea (ribbon worms; see text p56)

Phylum Gnathostomulida (gnathostomulids; see Further Information—*planktonic and interstitial forms* p133)

Phylum Loricifera (loriciferans; see Further Information—*planktonic and interstitial forms* p133)

Phylum Cycliophora (cycliophorans; see Further Information—*planktonic and interstitial forms* p133)

Phylum Sipunculida (acorn worms; see text p105)

Phylum Annelida (including Phylum Echiura; segmented worms, leeches; see text p58)

Phylum Entoprocta (kamptozoans or nodding heads; see text p60)

Phylum Phoronida (horseshoe worms; see text p60)

Phylum Nematoda (roundworms; see text p61)

Phylum Priapulida (penis worms; see text p61)

Phylum Tardigrada (water-bears; see Further Information—*planktonic and interstitial forms* p133)

Phylum Arthropoda (insects, spiders, crustaceans; see text p64)

- Subphylum Pycnognida (sea spiders; see text p66)
- Subphylum Hexapoda (insects; see text p64)
- Subphylum Cheliceriformes (spiders, scorpions, mites, horseshoe crabs)
  - Order Acari (mites; see text p65)
- Subphylum Crustacea (crustaceans; see text p69)
  - Subclass Copepoda (copepods; see text p70)
  - Class Ostracoda (ostracods; see text p70)
  - Class Malacostraca: (includes mantis shrimps, nebalians, krill, as well as the following)
    - Order Decapoda (crabs, prawns, lobsters; see text p76)
    - Order Isopoda (pill bugs, sea lice; see text p73)
    - Order Amphipoda (sand hoppers; see text p74)
    - Order Tanaidacea (tanaids; see text p71)
    - Order Cumacea (cumaceans; see text p71)
    - Order Mysida (mysids; see text p69)
  - Class Cirripedia (barnacles; see text p78)

Phylum Mollusca (clams, snails, squid; see text p81)

- Class Polyplacophora (chitons; see text p82)

Class Bivalvia (clams, mussels; see text p86)

Class Scaphopoda (tusk shells; see text p83)

Class Gastropoda (slugs and snails; see text p.84)

Class Aplacophora (aplacophorans; see text p83)

Class Monoplacophora (monoplacophorans; see Further Information—*molluscs* p145)

Class Cephalopoda (octopuses, cuttlefish, squid; see text p88)

Phylum Bryozoa (=Phylum Ectoprocta; lace corals; see text p102)

Phylum Brachiopoda (lamp shells; see text p104)

Phylum Echinodermata (sea stars and relatives; see text p91)

Class Crinoidea (feather stars; see text p93)

Class Asteroidea (sea stars; see text p95)

Class Ophiuroidea (brittle stars; see text p97)

Class Echinoidea (sea urchins; see text p98)

Class Holothuroidea (sea cucumbers; see text p100)

Phylum Hemichordata (acorn worms; see text p105)

Class Enteropneusta (acorn worms; see text p105)

Class Pterobranchia (pterobranchs; see text p105)

Phylum Chaetognatha (arrow worms; see text p105)

Phylum Chordata (sea squirts, salps, humans; see text p107)

Subphylum Cephalochordata (lancelets; see text p107)

Subphylum Urochordata

Class Ascidiacea (ascidians, sea squirts; see text p108)

Class Thaliacea (salps; see text p109)

Subphylum Vertebrata

Class Pteraspidomorphi (hagfishes; see Further Information—*Vertebrates* p151)

Class Cephalaspidomorphi (lampreys; see Further Information—*Vertebrates* p151)

Class Chondrichthyes (chimaeras; see text p111, sharks and rays; see text p112)

Class Pisces (bony fishes; see text p114)

Class Reptilia (crocodiles, turtles, sea snakes; see text p116)

Class Aves (birds; see text p117)

Class Mammalia (seals, dolphins, humans and relatives; see text p121)

GLOSSARY

asexual reproduction new individuals formed by buds or fragments from a single parent individual

benthic living on the sea floor

bilateral symmetry form based on linear pattern with right and left hand paired structures

bioluminescence generation of light by a living organism through chemical reaction

calcite crystals of calcium carbonate, occurring as a widely found inorganic mineral, and also forming the external shell or skeleton of some algae and many kinds of invertebrate

cartilage dense, more or less flexible connective tissue made from a protein called collagen and forming the skeleton of sharks and rays, and also found in some molluscs and other invertebrates

chaeta bristle protruding from the body of polychaete worms (termed 'seta' in many publications, however **seta** should be reserved for the distinct structures found in arthropods)

cilia minute hair-like structures found on the feeding tentacles of polychaetes, bryozoans, brachiopods, and several other kinds of invertebrate

Class rank below **Phylum** in the classification of living things; each **Class** is a group of related **Orders**

colonial many individual organisms living joined together as a colony

coralline with a hard shell or crust made of calcium carbonate

deposit feeding obtaining food from organic matter deposited on or in sediments

elasmobranch another name for members of the **Class** Chondrichthyes (sharks & rays)

endemic restricted to a particular geographical area, for example southern Australia

Family rank below **Order** in the classification of living things; each **Family** is a group of related **Genera**

fertilisation joining of genetic material from egg and sperm to form a new organism

filter feeding obtaining food by filtering particles from the water

foot muscular structure found in molluscs, often including a flattened sole on which the snail or other mollusc moves about

Genus rank below **Family** in the classification of living things; each **Genus** is a group of related species (genus names always written in **italic** font)

herbivorous feeding on plants or algae

hermaphrodite with both male and female sexual organs

holdfast root- or sucker-like structure that attaches brown seaweeds to rocks

intertidal occuring between high and low tide marks

introvert extensible mouthparts of sipunculan worms

Kingdom highest rank in the classification of living things; each **Kingdom** is a group of related **Phyla**

lophophore coil or crescent of tentacles bearing **cilia** found in bryozoans and brachiopods, used for **filter feeding**

maritime habitat within reach of the spray zone at and above high tide level

medusa floating jellyfish-like body form, often disc- or umbrella-shaped

nematocysts stinging cells found only in Cnidaria and used to catch prey

notochord flexible rod supporting the spinal nerve chord, present during the larval development of all chordates

Order rank below **Class** in the classification of living things; each **Order** is a group of related **Families**

ossicle minute rod- or barb-shaped skeletal element, made of calcite, found in sea cucumbers

pharynx name given to the muscular feeding structure of flatworms and polychaetes

photosynthesis process through which plants and algae store the energy of the sun as the sugar glucose

Phylum rank below **Kingdom** in the classification of living things; each **Phylum** is a group of related **Classes**

pinnules name given to the repeated elements forming a comb-like **filter feeding** tentacles in crinoids and in some polychaete worms

planktonic free swimming in the water away from the sea floor

planula free-swimming larval stage in life history of cnidarians

pneumatophores vertical extension providing air to the buried roots of mangroves

polyp flower-like body form with central opening surrounded by tentacles

proboscis name given to the extensible mouthparts of nemertean worms, of pygnogonids, and of acorn worms

radial symmetry form based on regular star- or wheel-like pattern with radial structures

rhizomes roots of seagrasses, buried in the sediment

scyphostoma tiny bottom-dwelling stage in the life history of scyphozoan jellyfishes

segmented formed from multiple similar units in a repeating pattern.

sessile fixed to a substrate

seta bristle protruding from the exoskeleton of crustaceans arthropods (see also **chaeta**)

sexual reproduction forming new individuals through fertilisation of eggs and sperm from parents of different sex

siphon tubes through which molluscs and ascidians move water, for filter feeding or for propulsion while swimming

solitary a single, separate individual organism

species rank below **Genus** in the classification of living things; each species is a group of related individuals capable of interbreeding and unable to reproduce with individuals belonging to other species (but often the subject of discussion among taxonomists and many definitions have been proposed; species names are always written in **italic** font)

stipe stalk of a brown seaweed

subtidal occuring below low tide mark

taxonomist student of taxonomy, the science of recognising, naming and classifying living things

tube feet fluid filled tentacle-like feet found only in echinoderms

valve in bivalved molluscs and brachiopods, the name given to each of the two shells

vesicles gas-filled floats found in some seaweeds

zooid individual organism from which colonies of bryozoans and kamptozoans are formed

FURTHER INFORMATION

The following references, web sites, and other resources are provided as a guide for those who need further Information. Our goal has been, for all marine life (even organisms not covered well in the body of the text) to provide an introduction to the literature, emphasising recent and readily available popular publications where possible. However, published knowledge of marine biodiversity, especially for Australia, is very uneven and incomplete. If we know of no alternative sources then we have listed technical academic publications, even though these will probably only be accessible through a university library.

The arrangement is taxonomic, after the first five headings (up to and including *Interstitial and planktonic forms*). The scope is decidedly 'southern Australia', but also included are some comprehensive works relevant to the New Zealand fauna, or to that of tropical Australia. The following list is also available from the web pages supporting this book: http://researchdata.museum.vic.gov.au/marine

Dangerous marine life and first aid treatment

Coleman, N. 2001. ***Dangerous Sea Creatures: An Aquatic Survival Guide.*** Atoll Editions. 96pp.

Edmonds, C. 1989. ***Dangerous Marine Creatures.*** Reed Books: Frenchs Forest. 192pp.

Fenner, P. 2007. ***Marine-medic.com.*** http://www.marine-medic.com.au/

Marsh, L.M., and Gurry, D.L. 1986. ***Sea Stingers - and other venomous and poisonous marine invertebrates of Western Australia.*** Western Australian Museum: Perth. 133pp.

Surf Lifesaving Queensland. 2007. ***Dangerous Marine Life.*** http://www.lifesaving.com.au/beachSmart/dangerousMarineLife.cfm.

Underhill, D. 1995. ***Australia's Dangerous Creatures.*** First Edition, Fourth Revised edn. Reader's Digest: Surry Hills. 368pp.

General information about marine life

Australian Biological Resources Study, 2006. *Australian Faunal Directory.* http://www.deh.gov.au/biodiversity/abrs/online-resources/fauna/afd/

Australian Museum Online, 2006. *Wildlife of Sydney.* http://www.faunanet.gov.au/wos/

Brusca, R.C. and Brusca, G.J. 2003. *Invertebrates.* Sinauer Associates: Sunderland, Massachusetts. xx & 936pp.

Dakin, W.J., and Bennett, I. 1987. *Australian Seashores.* Angus & Robertson: North Ryde, New South Wales.

Edgar, G.J. 2000. *Australian Marine Life*. revised edn. Reed: Kew, Victoria. 544pp.

Edgar, G.J. 2001. *Australian Marine Habitats in Temperate Waters*. Reed New Holland: Sydney. 224pp.

Lincoln, R.J., and Sheals, J.G. 1979. *Invertebrate Animals: Collection and Preservation*. British Museum (Natural History) and Cambridge University Press: Cambridge. vii + 150pp.

Maddison, D.R., and Schulz, K.-S. 1996–2006. *The Tree of Life Web Project*. http://tolweb.org

Margulis, L., and Schwartz, K.V. 1998. *Five Kingdoms: an illustrated guide to the Phyla of life on Earth*. 3rd edn. WH Freeman and Company: San Francisco. xx + 520pp.

Phillips, D.A.B., Handreck, C.P., Bock, P.E., Burn, R., Smith, B.J., and Staples, D.A. (eds) 1984. *Coastal Invertebrates of Victoria: an atlas of selected species*. Marine Research Group of Victoria and Museum of Victoria: Melbourne. v + 168pp.

Shepherd, S.A., and Davies, M. 1997. *Marine Invertebrates of Southern Australia Part III*. Handbooks Committee of the South Australian Government: Adelaide. 901–1264pp.

Shepherd, S.A., and Thomas, I.M. 1982. *Marine Invertebrates of Southern Australia Part I*. Handbooks Committee of the South Australian Government: Adelaide. 491pp.

Shepherd, S.A., and Thomas, I.M. 1989. *Marine Invertebrates of Southern Australia Part II*. Handbooks Committee of the South Australian Government: Adelaide. 497–900pp.

Societies

ANGAIR–Anglesea and Airey's Inlet Society for the Protection of Flora and Fauna. http://users.pipeline.com.au/angair/

Australian Marine Conservation Society. www.amcs.org.au/

Field Naturalists Club of Victoria. http://home.vicnet.net.au/~fncv/

Field Naturalists Society of SA. www.parks.sa.gov.au/parks/friends/groups/field_naturalists/

Malacalogical Society of Australasia. www.malsocaus.org

The Marine and Coastal Community Network. http://www.mccn.org.au/

Marine Research Group. http://home.vicnet.net.au/~fncv/marine.htm

ReefWatch Victoria. http://reefwatchvic.asn.au/

South-east Australian Naturalists Association. http://home.vicnet.net.au/~seana/

The Queensland Naturalists Club. http://www.qnc.org.au/

Microbes, protists

Ajani, P., Hallegraeff, G.M., and Pritchard, T. 2001. Historic overview of algal blooms in marine and estuarine waters of New South Wales, Australia. *Proceedings of the Linnean Society of New South Wales* 123: 1–22.

Albani, A.D. 1979. Recent shallow water Foraminiferida from New South Wales. *Australian Marine Sciences Association Handbook* 3: 1–51.

Collins, A.C. 1974. Port Phillip Bay Survey Foraminiferida. *Memoirs of the National Museum of Victoria* 35: 1–61, plates 61–64.

Darnton, B., and van Egmond, W. 1999–2006. *Foram Gallery*. http://www.microscopy-uk.org.uk/mag/indexmag.htmlhttp://www.microscopy-uk.org.uk/mag/artmar00/forwim.html

Darnton, B., and van Egmond, W. 1999–2006. *Radiolaria –Spectacular objects for the microscope*. http://www.microscopyuk.org.uk/mag/indexmag.htmlhttp://www.microscopy-uk.org.uk/mag/artjun99/bdradio.html

Hallegraeff, G.M., Anderson, D.M., and Cembella, A.D. 1995. *Manual on Harmful Marine Microalgae*. http://unesdoc.unesco.org/images/0012/001220/122021e0.pdf.

Hallegraeff, G.M., Anderson, D.M., and Cembella, A.D. (eds) 2003. *Manual on Harmful Marine Microalgae*. 2nd edn. UNESCO Publishing: Paris. 794pp.

Hayward, B.W., Grenfell, H.R., Reid, C.M., and Hayward, K.A. 1999. Recent New Zealand shallow-water benthic foraminifera: taxonomy, ecologic distribution, biogeography, and use in paleoenvironmental assessment. *New Zealand Geological Survey Institute of Geological & Nuclear Sciences Monograph* 21: 1–264.

Howey, R.L., and van Egmond, W. 1999–2006. *Welcome to the Wonderfully Weird World of Rotifers*. http://www.microscopyuk.org.uk/mag/indexmag.html?http://www.microscopy-uk.org.uk/mag/artnov99/rotih.html

Huisman, J., Matthijs, H.C.P., and Visser, P.M. (eds) 2005. *Harmful Cyanobacteria*. University of Western Australia Press: Perth. 243pp.

Patterson, D.J. 1996. *Free-living Freshwater Protozoa: A colour guide*. Manson Publishing: London. 223pp.

Scott, F.J., and Marchant, H.J. (eds) 2005. *Antarctic Marine Protists*. Australian Biological Resources Study and Australian Antarctic Division: Canberra and Hobart. 550pp.

Parasites

Rohde, K. (ed.) 2005. *Marine Parasitology*. CSIRO Publishing: Collingwood. 565pp.

Interstitial and planktonic forms

Bradford-Grieve, J.M. 2002. *Calanoida: Families*. http://www.crustacea.net/

d'Hondt, J.L. 1971. Gastrotricha. *Oceanography and Marine Biology, An Annual Review* 9: 141–192.

Hulings, N.C., and Gray, J.S. 1971. A manual for the study of Meiofauna. *Smithsonian Contributions to Zoology* 78: 1–83.

Hummon, W.D. 1974. Some taxonomic revisions and nomenclatural notes concerning marine and brackish-water Gastrotricha. *Transactions of the American Microscopical Society* 93: 194–205.

Johnsson, R., and Rocha, C.E.F. 2002. Five artotrogids (Crustacea: Copepoda: Siphonostomatoida) from Eastern Antarctica. *Memoirs of Museum Victoria* 59: 439–456.

Nicholas, W., and Todaro, M.A. 2006. Two new species of Tetranchyroderma (Gastrotricha, Macrodasyida) from a sandy beach in southeastern Australia. *New Zealand Journal of Marine and Freshwater Research* 40: 249–258.

O'Sullivan, D., and Hosie, G. 1985. A general guide to the metazoan zooplankton groups of the Southern Ocean. *ANARE Research Notes* 30: 1–59.

Ritz, D., Swadling, K.M., Hosie, G., and Cazassus, F. 2003. Guide to the Zooplankton of south eastern Australia. *Fauna of Tasmania Handbook no. 10*. Fauna of Tasmania Committee, University of Tasmania: Hobart. 91pp.

Somerfield, P.J., and Warwick, R.M. 1996. *Meiofauna in Marine Pollution Monitoring Programmes: A Laboratory Manual*. Ministry of Agriculture Fisheries and Food Directorate of Fisheries Research: Lowestoft. 71pp.

Thiel, H., and Higgins, R.P. 1988. *Introduction to the study of Meiofauna*. Smithsonian Institution Press: Washington, DC 488pp.

Walter, T.C. 2006. *World of Copepods*. http://www.nmnh.si.edu/iz/copepoda/

[See also Zeidler references on pelagic amphipoda, below, under *Sand hoppers (Subphylum Crustacea, Order Amphipoda)*]

Lichens

Filson, R.B., and Rogers, R.W. 1979. *Lichens of South Australia*. S.A. Government Printer: Adelaide. 197pp.

Martin, W., and Child, J. 1972. *New Zealand Lichens*. A.W. Reed: Wellington. 193pp.

Purvis, W. 2000. *Lichens*. Natural History Museum: London. 112pp.

Seaweeds, algae

Clayton, M.N., and King, R.J. (eds) 1990. *Biology of Marine Plants*. Longman: Melbourne. 501pp.

Fuhrer, B., Christianson, I.G., Clayton, M.N., and Allender, B.M. 1981. *Seaweeds of Australia*. A.H. & A.W. Reed: Sydney. 112pp.

Huisman, J. 2000. *Marine Plants of Australia*. University of Western Australia Press: Perth. 300pp.

Womersley, H.B.S. (ed.) 1984. *Marine Benthic Flora of Southern Australia–Part I*. Government Printer: Adelaide. 329pp.

Womersley, H.B.S. 1987. *Marine Benthic Flora of Southern Australia–Part II*. Government Printer: Adelaide. 481pp.

Womersley, H.B.S. 1994. *Marine Benthic Flora of Southern Australia–Part IIIA–Bangiophyceae and Florideophyceae (Acrochaetiales, Nemaliales, Gelidiales, Hildenbrandiales and Gigartinales sensu lato)*. Australian Biological Resources Study: Canberra. 508pp.

Womersley, H.B.S. 1996. *Marine Benthic Flora of Southern Australi a– Part IIIB–Gracilariales, Rhodymeniales, Corallinales and Bonnemaisoniales*. Australian Biological Resources Study: Canberra.

Womersley, H.B.S. 1998. *Marine Benthic Flora of Southern Australia–Part IIIC–Ceramiales–Ceramiaceae, Dasyaceae*. Australian Biological Resources Study: Canberra. 535pp.

Womersley, H.B.S. 2003. *Marine Benthic Flora of Southern Australia–Part IIID Ceramiales–Delesseriaceae, Sarcomeniaceae, Rhodomelaceae*. Australian Biological Resources Study: Canberra. 533pp.

Seagrasses, mangroves (Angiosperms)

Clayton, M.N., and King, R.J. (eds) 1990. *Biology of Marine Plants.* Longman: Melbourne. 501pp.

Costermans, L. 1983. *Native Trees and Shrubs of South-eastern Australia.* Rigby: Adelaide. 422pp.

Kuo, J., and McComb, A.J. 1989. Seagrass taxonomy, structure and development. Pp. 6–73 in: Larkum, A.W.D., McComb, A.J., and Shepherd, S.A. (eds) *Biology of Seagrasses: A Treatise on the Biology of Seagrasses with Special Reference to the Australian Region.* Elsevier: Amsterdam.

Robertson, E.L. 1984. Seagrasses. Pp. 57–122 in: Womersley, H.B.S. (ed.) *Marine Benthic Flora of Southern Australia–Part I.* Government Printer: Adelaide.

Sponges (Phylum Porifera)

Bergquist, P.R., and Skinner, I.G. 1982. Sponges (Phylum Porifera). Pp. 38–72 in: Shepherd, S.A., and Thomas, I.M. (eds) *Marine Invertebrates of Southern Australia Part 1.* Handbooks Committee of the South Australian Government: Adelaide.

Hooper, J.N.A. 2000. *Sponguide: Guide to sponge collection and identification.* http://www.qmuseum.qld.gov.au/organisation/sections/SessileMarineInvertebrates/spong.pdf

Hooper, J.N.A., Soest, R.W.M.v., and Willenz, P. 2002. *Systema Porifera: A guide to the classification of sponges.* Springer: Berlin. 2 volumes, 1708pp.

Hooper, J.N.A., and Wiedenmayer, F. 1994. *Zoological Catalogue of Australia Volume 12. Porifera.* CSIRO Information Services: East Melbourne, Vic. 624pp.

Picton, B. 1997. *Porifera Web Page.* http://www.jiscmail.ac.uk/files/PORIFERA/welcome.html

Phylum Cnidaria–general

University of California Irvine. 2006. *The Cnidaria Home Page.* http://www.ucihs.uci.edu/biochem/steele/default.html

Hydroids (Phylum Cnidaria, Class Hydrozoa)

O'Sullivan, D. 1982. A guide to the Hydromedusae of the Southern Ocean and adjacent waters. *ANARE Research Notes* 5.

Schuchert, P. 1996. The Marine Fauna of New Zealand: Athecate Hydroids and their medusae (Cnidaria: Hydrozoa). *New Zealand Oceanographic Institute Memoir* 106: 1–159.

Schuchert, P. 2006. The Hydrozoa Directory. Version 11, April 2006. http://www.ville-ge.ch/musinfo/mhng/hydrozoa/hydrozoa-directory.htm

Vervoort, W., and Watson, J.E. 2003. The Marine Fauna of New Zealand: Leptothecata (Cnidaria: Hydrozoa) (Thecate Hydroids). *NIWA Biodiversity Memoir* 119: 1–538.

Watson, J.E. 1982. Hydroids (Class Hydrozoa). Pp. 77–114 in: Shepherd, S.A., and Thomas, I.M. (eds) *Marine Invertebrates of Southern Australia Part 1.* Handbooks Committee of the South Australian Government: Adelaide.

Watson, J.E. 2000. Hydroids (Hydrozoa: Leptothecatae) from the Beagle Gulf and Darwin Harbour, northern Australia. *Beagle* 16: 1–82.

Watson, J.E. 2002. *Corystolona*, a new hydroid genus (Leptolida: Leptothecate) from southern Australia. *Memoirs of Museum Victoria* 59: 333–336.

Sea anemones, corals (Phylum Cnidaria, Class Anthozoa)

Alderslade, P. 1998. Revisionary systematics in the gorgonian family Isididae, with descriptions of numerous new taxa (Coelenterata: Octocorallia). *Records of the Western Australian Museum Supplement* 55: 1–359.

Alderslade, P. 2003. A new genus and new species of soft coral (Octocorallia: Alcyonacea: Alcyoniidae) from the south western region of Australia. *Zootaxa* 175: 1–10.

Alderslade, P. 2006. New subfamilies and a new genus and species of Melithaeidae (Coelenterata: Octocorallia: Alcyonacea) with comparative data on the structure of both melithaeid and subergorgiid axes. *Zootaxa* 1199: 19–47.

Cutress, C.E. 1971. Corallimorpharia, Actiniaria and Zoanthidea. *Memoirs of the National Museum of Victoria* 32: 83–92, plate 89.

Fabricius, K.E., and Alderslade, P. 2001. *Soft Corals and Sea Fans*. Australian Institute of Marine Science: Townsville. viii & 264pp.

Fautin, D.G. 2006. *Hexacorallians of the world–sea anemones, corals and their allies*. http://hercules.kgs.ku.edu/hexacoral/anemone2

Grasshoff, M. 1982. Pseudogorgonian (Order Gastraxonacea). Pp. 195–197 in: Shepherd, S.A., and Thomas, I.M. (eds) *Marine Invertebrates of Southern Australia Part 1*. Handbooks Committee of the South Australian Government: Adelaide.

Grasshoff, M. 1982. Gorgonians or Sea Fans (Order Gorgonacea). Pp. 198–206 in: Shepherd, S.A., and Thomas, I.M. (eds) *Marine Invertebrates of Southern Australia Part 1*. Handbooks Committee of the South Australian Government: Adelaide.

Shepherd, S.A., and Veron, J.E.N. 1982. Stony Corals (Order Scleractinia or Madreporararia). Pp. 169–178 in: Shepherd, S.A., and Thomas, I.M. (eds) *Marine Invertebrates of Southern Australia Part 1*. Handbooks Committee of the South Australian Government: Adelaide.

Thomas, I.M., and Shepherd, S.A. 1982. Sea Anemones (Orders Actiniaria, Zoanthidea and Corallimorpharia). Pp. 160–168 in: Shepherd, S.A., and Thomas, I.M. (eds) *Marine Invertebrates of Southern Australia Part 1*. Handbooks Committee of the South Australian Government: Adelaide.

Utinomi, H. 1971. Octocorallia. *Memoirs of the National Museum of Victoria* 32: 7–18.

Utinomi, H., and Shepherd, S.A. 1982. Seapens (Order Pennatulacea). Pp. 207–219 in: Shepherd, S.A., and Thomas, I.M. (eds) *Marine Invertebrates of Southern Australia Part 1*. Handbooks Committee of the South Australian Government: Adelaide.

Veron, J.E.N. 1996. *Corals of the World*. Volume 1, 1996. Volume 2, 2000. Volume 3, 2003. Australian Institute of Marine Science: Townsville, Australia.

Veron, J.E.N., and Stafford-Smith, S. 2002. *Coral ID: An electronic guide to the zooxanthellate scleractinians of the world [CD-ROM]*. Australian Institute of Marine Science: Townsville, Australia.

Verseveldt, J. 1982. Soft Corals or Alcyonarians (Orders Stolonifera, Telestacea and Alcyonacea). Pp. 183–194 in: Shepherd, S.A., and Thomas, I.M. (eds) *Marine Invertebrates of Southern Australia Part 1*. Handbooks Committee of the South Australian Government: Adelaide.

Williams, G.C. 1995. Revision of the pennatulacean genus *Sarcoptilus* (Coelenterata: Octocorallia), with descriptions of three new species from southern Australia. *Records of the South Australian Museum* 28: 13–32

Jellyfish (Phylum Cnidaria, Class Scyphozoa)

Fancett, M.S. 1986. Species composition and abundance of Scyphomedusae in Port Phillip Bay, Victoria. *Australian Journal of Marine and Freshwater Research* 37: 379–384.

O'Sullivan, D. 1982. A guide to the Scyphomedusae of the Southern Ocean and adjacent waters. *ANARE Research Notes* 4.

Southcott, R.V. 1971. Medusae Coelenterata. *Memoirs of the National Museum of Victoria* 32: 1–6, plates 1–5.

Southcott, R.V. 1982. Jellyfishes (Classes Scyphozoa and Hydrozoa). Pp. 115–159 in: Shepherd, S.A., and Thomas, I.M. (eds) *Marine Invertebrates of Southern Australia Part 1*. Handbooks Committee of the South Australian Government: Adelaide.

Wrobel, D. 2006. The Jellie Zone. http://jellieszone.com/

Wrobel, D., and Mills, C. 1998. *Pacific Coast Pelagic Invertebrates*. Sea Challengers: Monterey. 108pp.

Box jellies (Phylum Cnidaria, Class Cubozoa)

Matsumoto, G.I. 1995. Observations on the anatomy and behaviour of the cubozoan *Carybdea rastonii* Haacke. *Marine and Freshwater Behaviour and Physiology* 26: 139–148.

Southcott, R.V. 1982. Jellyfishes (Classes Scyphozoa and Hydrozoa). Pp. 115–159 in: Shepherd, S.A., and Thomas, I.M. (eds) *Marine Invertebrates of Southern Australia Part 1*. Handbooks Committee of the South Australian Government: Adelaide.

[see also Wrobel publications under *Jellyfish (Phylum Cnidaria, Class Schypozoa)*]

Comb jellies (Phylum Ctenophora)

Harbison, G.R. 1985. On the classification and evolution of the Ctenophora. Pp. 78–99 in: Conway Morris, S., George, J.D., Gibson, R., and Platt, H.M. (eds) *The Origins and Relationships of Lower Invertebrates*. The Systematics Association and Clarendon Press: Oxford.

Harbison, G.R., and Madin, L.P. 1982. Ctenophora. Pp. 707–715 in: Parker, S.P. (ed.) *Synopsis and Classification of Living Organisms*. McGraw-Hill: New York.

Matsumoto, G.I. 1999. *Coeloplana thomsoni* sp. nov., a new benthic ctenophore (Ctenophora: Platyctenida: Coeloplanidae) from Western Australia. Pp. 385–394 in: Walker, I., and Wells, F.E. (eds) *The Seagrass Flora and Fauna of Rottnest Island, Western Australia, Perth*. Western Australian Museum: Perth.

Matsumoto, G.I., and Gowlett-Holmes, K. 1996. *Coeloplana scaberi* sp. nov., a new benthic ctenophore (Phylum Ctenophora, Order Platyctenida) from South Australia. *Records of the South Australian Museum* 29: 33–40.

Mills, C.E. 1998–2006. Phylum Ctenophora: list of all valid species names. http://faculty.washington.edu/cemills/Ctenolist.html

Smith, B.J., and Plant, R.J. 1976. A creeping ctenophoran (Platyctenea: Ctenophora) from Victoria, Australia. *Memoirs of the National Museum of Victoria* 37: 43–46.

van Egmond, W. 1999–2006. *Comb Jellies*. http://www.microscopyuk.org.uk/mag/artmay98/comb.html

[see also Wrobel publications under *Jellyfish (Phylum Cnidaria, Class Schypozoa)*]

Flatworms (Phylum Platyhelminthes)

Cannon, L.R.G. 1986. *Turbellaria of the World. A Guide to Families and Genera.* Queensland Museum: Brisbane. vii + 136pp.

Lee, K.M., Beal, M.A., and Johnston, E.L. 2006. A new predatory flatworm (Platyhelminthes, Polycladida) from Botany Bay, New South Wales, Australia. *Journal of Natural History* 39: 3987–3995.

Newman, L.J., and Cannon, L.R.G. 2003. *Marine Flatworms: The world of polyclads.* CSIRO Publishing: Melbourne. 112pp.

Newman, L.J., and Cannon, L.R.G. 2005. *Fabulous Flatworms: A guide to marine polyclads [CD-ROM].* CSIRO Publishing: Melbourne.

Prudhoe, S. 1978. Some polyclad turbellarians new to the fauna of the Australian coasts. *Records of the Australian Museum* 31: 586–604.

Prudhoe, S. 1982. Polyclad flatworms (Phylum Platyhelminthes). Pp. 220–227 in: Shepherd, S.A., and Thomas, I.M. (eds) *Marine Invertebrates of Southern Australia Part 1.* Handbooks Committee of the South Australian Government: Adelaide.

Prudhoe, S. 1982. Polyclad Turbellarians of the southern coasts of Australia. *Records of the South Australian Museum* 18: 361–384.

Ribbon worms (Phylum Nemertea)

Gibson, R. 1972. *Nemerteans.* Hutchinson University Library: London. 224pp.

Gibson, R. 1990. The macrobenthic nemertean fauna of the Albany region, Western Australia. Pp. 89–194 in: Wells, F.E., Walker, D.I., Kirkman, H., and Lethbridge, R. (eds) *Proceedings of the Third International Marine Biological Workshop: The marine flora and fauna of Albany, Western Australia.* Western Australian Museum: Perth.

Gibson, R. 1990. A new species of Carcinonemertes (Nemertea: Enopla: Carcinonemertidae) from the egg masses of Naxia auita (Latreille) (Decapoda: Brachyura: Majidae) collected in the Albany region of Western Australia. Pp. 195–202 in: Wells, F.E., Walker, D.I., Kirkman, H., and Lethbridge, R. (eds) *Proceedings of the Third International Marine Biological Workshop: The marine flora and fauna of Albany, Western Australia.* Western Australian Museum: Perth.

Gibson, R. 1995. Nemertean genera and species of the world: an annotated checklist of original names and description citations, synonyms, current taxonomic status, habitats and recorded zoogeographic distribution. *Journal of Natural History* 29: 271–562.

Gibson, R. 1997. Nemerteans (Phylum Nemertea). Pp. 905–974 in: Shepherd, S.A., and Davies, M. (eds) *Marine Invertebrates of Southern Australia Part III.* Handbooks Committee of the South Australian Government: Adelaide.

Gibson, R. 1999. Further studies on the nemertean fauna of Rottnest Island, Western Australia. Pp. 359–376 in: Walker, I., and Wells, F.E. (eds) *The Seagrass Flora and Fauna of Rottnest Island, Western Australia, Perth.* Western Australian Museum: Perth.

Gibson, R. 2002. The invertebrate fauna of New Zealand: Nemetea (ribbon worms). *NIWA Biodiversity Memoir* 118: 1–87.

Norenburg, J. 2006. *Nemertes Portal.* http://nemertes.si.edu

Sundberg, P., and Gibson, R. 1995. The nemerteans (Nemertea) of Rottnest Island, Western Australia. *Zoologica Scripta* 24: 101–141.

Peanut worms (Phylum Sipuncula)

Cutler, E.B., Schulz, A., and Dean, H.K. 2004. The Sipuncula of sublittoral New Zealand, with a key to all New Zealand species. *Zootaxa* 525: 1–19.

Edmonds, S.J. 1980. A revision of the systematics of Australian sipunculans (Sipuncula). *Records of the Australian Museum* 18: 1–74.

Edmonds, S.J. 1982. Sipunculans (Phylum Sipuncula). Pp. 299–311 in: Shepherd, S.A., and Thomas, I.M. (eds) *Marine Invertebrates of Southern Australia Part 1*. Handbooks Committee of the South Australian Government: Adelaide.

Edmonds, S.J. 2000. Phylum Sipuncula. Pp. 375–400 in: Beesley, P.L., Ross, G.J.B., and Glasby, C.J. (eds) *Polychaetes and Allies: The southern synthesis*. CSIRO Publishing: Melbourne.

Stephen, A.C., and Edmonds, S.J. 1972. *The Phyla Sipuncula and Echiura*. Trustees of the British Museum (Natural History): London. vii + 529pp.

Segmented worms (polychaetes, oligochaetes, leeches Phylum Annelida; including Echiura)

Beesley, P.L., Ross, G.J.B., and Glasby, C.J. (eds) 2000. *Polychaetes & Allies: The southern synthesis. Fauna of Australia Vol 4A*. CSIRO Publishing: Melbourne. xii + 465pp.

Edmonds, S.J. 1982. Echiurans (Phylum Echiura). Pp. 312–318 in: Shepherd, S.A., and Thomas, I.M. (eds) *Marine Invertebrates of Southern Australia Part 1*. Handbooks Committee of the South Australian Government: Adelaide.

Edmonds, S.J. 1987. Echiurans from Australia (Echiura). *Records of the South Australian Museum* 32: 119–138.

Edmonds, S.J. 2000. Phylum Echiura. Pp. 353–374 in: Beesley, P.L., Ross, G.J.B., and Glasby, C.J. (eds) *Polychaetes and Allies: The southern synthesis*. CSIRO Publishing: Melbourne.

Meyer, M.C., and Burreson, E.M. 1990. Some leeches (Hirudinea: Piscicolidae) of the southern oceans. *Antarctic Research Series* 52: 219–236.

O'Sullivan, D. 1982. A guide to the pelagic polychaetes of the Southern Ocean and adjacent waters. *ANARE Research Notes* 3: 1–62.

Read, G.B. 1996–2006. *Annelid Resources*. http://www.annelida.net/

Read, G.B. 2004. *Guide to New Zealand Shore Polychaetes. Web publication*. http://www.niwa.cri.nz/ncabb/tools/index.html#mguides

Read, G.B. 2004. *Guide to New Zealand Shell Polychaetes. Web publication*. http://www.niwa.cri.nz/ncabb/tools/index.html#mguides

Richardson, L.R. 1959. N. Z. Hirudinea-IV. *Makarabdella manteri* n. g., n. sp., a new marine piscicolid leech. *Transactions of the Royal Society of New Zealand* 87: 283–290.

Richardson, L.R., and Meyer, M.C. 1973. Deep-sea fish leeches. *Galathea Report* 12: 113–126.

Wilson, R.S., Hutchings, P.A., and Glasby, C.J. 2003. *Polychaetes: An interactive identification guide [CD-ROM]*. CSIRO Publishing: Melbourne.

Phoronids (Phylum Phoronida)

Emig, C.-C., Boesch, D.F., and Rainer, S.F. 1977. Phoronida from Australia. *Records of the Australian Museum* 30: 455–474.

Emig, C.-C., and Roldan, C. 1992. The occurrence in Australia of three species of Phoronida (Lophophorata) and their distribution in the Pacific area. *Records of the South Australian Museum* 26: 1–8.

Shepherd, S.A. 1997. Phoronids (Phylum Phoronida). Pp. 993–999 in: Shepherd, S.A., and Davies, M. (eds) *Marine Invertebrates of Southern Australia Part III*. Handbooks Committee of the South Australian Government: Adelaide.

Kamptozoans, or nodding heads (Phylum Entoprocta)

Wasson, K. 2002. A review of the invertebrate phylum Kamptozoa (Entoprocta) and synopsis of kamptozoan diversity in Australia and New Zealand. *Transactions of the Royal Society of South Australia* 126: 1–20.

Wasson, K., and Shepherd, S.A. 1997. Nodding heads (Phylum Kamptozoa or Entoprocta). Pp. 975–992 in: Shepherd, S.A., and Davies, M. (eds) *Marine Invertebrates of Southern Australia Part III*. Handbooks Committee of the South Australian Government: Adelaide.

Roundworms (Phylum Nematoda)

Platt, H.M., and Warwick, R.M. 1983. *Free-living Marine Nematodes. Part I: British Enoplids. Synopses of the British Fauna (New series) No. 28*. Cambridge University Press: Cambridge. 307pp.

Platt, H.M., and Warwick, R.M. 1988. *Free-living Marine Nematodes. Part II: British Chromadorids. Synopses of the British Fauna (New series) No. 38*. Field Studies Council: Cambridge. 502pp.

Platt, H.M., and Warwick, R.M. 1998. *Free-living Marine Nematodes. Part III: British Monhysterids. Synopses of the British Fauna (New series) No. 53*. Field Studies Council: Cambridge. 296pp.

Riemann, F. 1988. Chapter 23. Nematoda. in: Higgins, R.P., and Thiel, H. (eds) *Introduction to the Study of Meiofauna*. Smithsonian Institution Press: Washington DC.

Steyaert, M. 2006. *Nemsys: Key to the free-living marine Nematoda*. http://intramar.ugent.be/nemys/ident/iden.asp?t=16390&k=13

University of Florida Institute of Food and Agricultural Sciences. 2006. *Illustrated Key to the Genera of Free-living Marine Nematodes in the Superfamily Chromadoroidea–Exclusive of the Chromadoridae*. http://creatures.ifas.ufl.edu/nematode/marine_nematodes.htm

Penis worms (Phylum Priapula)

Land, J.v.d. 1970. Systematics, zoogeography, and ecology of the Priapulida. *Zoologische Verhandelingen uitgegeven door het Rijksmuseum van Natuurlijke Historie te Leiden* 112: 1–118.

Land, J.v.d., and Nørrevang, A. 1985. Relationships of the Priapulida. Pp. 261–273 in: Conway Morris, S., George, J.D., Gibson, R., and Platt, H.M. (eds) *The Origins and Relationships of Lower Invertebrates*. The Systematics Association and Clarendon Press: Oxford.

Water bears (Phylum Tardigrada)

Dezio, S.G., Gallo, M.D., and Delucia, M.R.M. 1987. Adaptive radiation and phylogenesis in marine Tardigrada and the establishment of Neostygarctidae, a new family of Heterotardigrada. *Bollettino di Zoologia* 54: 27–33.

Kristensen, R.M., and Neuhaus, B. 1999. The ultrastructure of the tardigrade cuticle with special attention to marine species. *Zoologischer Anzeiger* 238: 261–281.

Noda, H. 1998. *Marine Tardigrada.* http://homepage3.nifty.com/cxj11255/tardigrada/index.html

Sea spiders (Phylum Arthropoda, subphylum Pycnogonida)

Child, C.A. 1975. Pycnogonida of Western Australia. *Smithsonian Contributions to Zoology* 190: 1–29.

Child, C.A. 1998. The marine fauna of New Zealand, Pycnogonida (Sea Spiders). *National Institute of Water and Atmospheric Research Biodiversity Memoir* 109: 4–71.

Clark, W.C. 1963. Australian Pycnogonida. Records of the Australian Museum 26: 1–81.

Staples, D.A. 1997. Sea spiders or pycnogonids (Phylum Arthropoda). Pp. 1040–1072 in: Shepherd, S.A., and Davies, M. (eds) *Marine Invertebrates of Southern Australia Part III.* Handbooks Committee of the South Australian Government: Adelaide.

Staples, D.A. 2002. *Pycnogonum* (Pycnogonida: Pycnogonidae) from Australia with descriptions of two new species. *Memoirs of Museum Victoria* 59: 541–553.

Staples, D.A., and Watson, J.E. 1987. Associations between pycnogonids and hydroids. Pp. 215–226 in: Bouillion, J. (ed.) *Modern Trends in the Systematics, Ecology, and Evolution of Hydroids and Hydromedusa.* Oxford University Press: Oxford.

Stock, J.H. 1973. *Achelia shepherdi* n. sp. and other Pycnogonida from Australia. *Beaufortia* 21: 91–97.

Stock, J.H. 1973. Pycnogonida from south-eastern Australia. *Beaufortia* 20: 99–127.

Insects (Phylum Arthropoda, subphylum Hexapoda)

Cranston, P.S. 1996. *Identification Guide to the Chironomidae of New South Wales.* Australian Water Technologies Pty Ltd: West Ryde, New South Wales, Australia. vii + 376pp.

CSIRO (Australia). Division of Entomology. 1991. *The Insects of Australia: A textbook for students and research workers.* 2nd edn. Melbourne University Press: Carlton South, Vic. xvi, 1137pp. (2 volumes)

Mites (Phylum Arthropoda, subphylum Cheliceriformes, order Acari)

Bartsch, I. 1993. Rhombognathine mites (Halacaridae, Acari) from Rottnest Island, Western Australia. Pp. 19–43 in: Wells, F.E., Walker, D., Kirkman, H., and Lethbridge, R. (eds) *Proceedings of the Fifth International Marine Biological Workshop: The marine flora and fauna of Rottnest Island, Western Australia.* Western Australian Museum: Perth.

Bartsch, I. 1993. *Halacarus* (Halacaridae, Acari) from south-western Australia. Pp. 45–71 in: Wells, F.E., Walker, D., Kirkman, H., and Lethbridge, R. (eds) *Proceedings of the Fifth International Marine Biological Workshop: The Marine Flora and Fauna of Rottnest Island, Western Australia.* Western Australian Museum: Perth.

Bartsch, I. 1993. Arenicolous Halacaridae (Acari) from South-Western Australia. Pp. 73–103 in: Wells, F.E., Walker, D., Kirkman, H., and Lethbridge, R. (eds) *Proceedings of the Fifth International Marine Biological Workshop: The Marine Flora and Fauna of Rottnest Island, Western Australia*. Western Australian Museum: Perth.

Bartsch, I. 1994. The genus *Simognathus* (Acari : Halacaridae), description of six new species from southern Australia and a tabular key to all species. *Acarologia* 35: 135–152.

Bartsch, I. 1994. A new species of the *Copidognathus tricorneatus* Group (Acari: Halacaridae) from Western Australia with a review of this species-group. *Species Diversity* 2: 155–166.

Bartsch, I. 1994. A new species of the *Copidognathus pulcher* group (Acari: Halacaridae) from Western Australia: Description of adults and juveniles and notes on developmental pattern. *Species Diversity* 3: 187–200.

Bartsch, I. 1994. *Arhodeoporus* (Acari: Halacaridae) from Rottnest Island, description of three new species. *Acarologia* 38: 265–274.

Bartsch, I. 1994. *Copidognathus* (Halacaridae: Acari) from Western Australia. Description of twelve species of the *gibbus* group. *Records of the Western Australian Museum* 16: 535–566.

Bartsch, I. 1996. Halacarines (Acari: Halacaridae) from Rottnest Island, Western Australia: the genera *Agauopsis* Viets and *Halacaropsis* gen. nov. *Records of the Western Australian Museum* 18: 1–18.

Bartsch, I. 1999. Halacaridae (Acari) from Rottnest Island: Description of two Agaue species. *Acarologia* 40: 179–190.

Bartsch, I. 1999. Halacaridae (Acari) from Western Australia. Four species of *Copidognathus*. Pp. 315–331 in: Walker, I., and Wells, F.E. (eds) *The Seagrass Flora and Fauna of Rottnest Island, Western Australia, Perth*. Western Australian Museum: Perth.

Bartsch, I. 1999. Halacaridae (Acari) from Rottnest Island, Western Australia. Mites on fronds of the seagrass *Amphibolis*. Pp. 333–357 in: Walker, I., and Wells, F.E. (eds) *The Seagrass Flora and Fauna of Rottnest Island, Western Australia, Perth*. Western Australian Museum: Perth.

Hunt, G.S. 1994. Orabatids: a mite biodiverse (Acarina). *Memoirs of the Queensland Museum* 36: 107–114.

Womersley, H.B.S. 1937. A new species of marine Hydrachnellae from South Australia. *Transactions of the Royal Society of South Australia* 61: 173–174.

Crustaceans (Subphylum Crustacea)—general

Ahyong, S.T., and Lowry, J.K. 2001. *Stomatopoda: Families*. http://www.crustacea.net/

Gerken, S. 2001. The Gynodiastylidae (Crustacea: Cumacea). *Memoirs of Museum Victoria* 59: 1–276.

Jones, D.S., and Morgan, G.J. 2002. *A Field Guide to Crustaceans of Australian Waters*. 2nd edn. Reed New Holland: Sydney. 224pp.

Larsen, K. 2002. *Tanaidacea: Families*. http://www.crustacea.net/

Martin, J.W., and Davis, G.E. 2001. An updated classification of the Recent Crustacea. *Natural History Museum of Los Angeles County, Science Series* 39: 1–124.

Meland, K. 2002. *Mysicacea: Families, Subfamilies and Tribes*. http://www.crustacea.net/

Shields, J. 2006. *The Crustacean Society.* http://vims.edu/tcs/

Walker-Smith, G.K. 1998. A review of *Nebaliella* (Crustacea: Leptostraca) with description of a new species from the continental slope of southeastern Australia. *Memoirs of the Museum of Victoria* 57: 39–56.

Walker-Smith, G.K. 2000. *Levinebalia maria*, a new genus and new species of Leptostraca (Crustacea) from Australia. *Memoirs of the Museum of Victoria* 58: 137–147.

Walker-Smith, G.K., and Poore, G.C.B. 2001. A phylogeny of the *Leptostraca* (Crustacea) with keys to the families and genera. *Memoirs of the Museum of Victoria* 58: 383–410.

Crabs, lobsters (Subphylum Crustacea, Order Decapoda)

McLaughlin, P., Ahyong, S.T., and Lowry, J.K. 2002. *Anomura: Families.* http://www.crustacea.net/.

Poore, G.C.B. 2004. *Marine Decapod Crustacea of Southern Australia: A Guide to Identification.* CSIRO Publishing: Melbourne. ix & 574pp.

Pill bugs, sea lice (Subphylum Crustacea, Order Isopoda)

Bruce, N.L. 2003. New genera and species of sphaeromatid isopod crustaceans from Australian marine coastal waters. *Memoirs of Museum Victoria* 60: 309–370.

Cohen, B.F., and Poore, G.C.B. 1994. Phylogeny and biogeography of the Gnathiidae (Crustacea: Isopoda) with descriptions of new genera and species, most from south-eastern Australia. *Memoirs of the Museum of Victoria* 54: 271–397.

Just, J., and Wilson, G.D.F. 2006. Revision of Southern Hemisphere *Austronanus* Hodgson, 1910, with two new genera and five new species of Paramunnidae (Crustacea: Isopoda: Asellota). *Zootaxa* 1111: 21–58.

Keable, S.J., Poore, G.C.B., and Wilson, G.D.F. 2002. *Isopoda: Families.* http://www.crustacea.net/

King, R.A. 2003. *Neastacilla* Tattersall, 1921 redefined, with eight new species from Australia (Crustacea: Isopoda: Arcturidae). *Memoirs of Museum Victoria* 60: 371–416.

Poore, G.C.B. 2001. Families and genera of Isopoda Anthuridea. *Crustacean Issues* 13: 63–173.

Poore, G.C.B., and Bardsley, T.M. 2004. Pseudidotheidae (Crustacea: Isopoda: Valvifera) reviewed with description of a new species, first from Australia. *Memoirs of Museum Victoria* 61: 75–83.

Poore, G.C.B., and Lew Ton, H.M. 1993. Idoteidae of Australia and New Zealand (Crustacea: Isopoda: Valvifera). *Invertebrate Taxonomy* 7: 197–278.

Poore, G.C.B., and Lew Ton, H.M. 2002. Expanathuridae (Crustacea: Isopoda) from the Australian region. *Zootaxa* 82: 1–60.

Sand hoppers (Subphylum Crustacea, Order Amphipoda)

Barnard, J.L., and Karaman, G.S. 1991. The families and genera of marine gammaridean Amphipoda (except marine gammaroids). Parts 1 and 2. *Records of the Australian Museum, Supplement* 13: 1–866.

Lowry, J.K., and Springthorpe, R.T. 2001. *Amphipoda: Families.* http://www.crustacea.net/

Lowry, J.K. 2006. New families and subfamilies of amphipod crustaceans. *Zootaxa* 1254: 1-28.

Zeidler, W. 2003. A review of the hyperiidean amphipod family Cystisomatidae Willemöes-Suhm, 1875 (Crustacea: Amphipoda: Hyperiidea). *Zootaxa* 141: 1–43.

Zeidler, W. 2003. A review of the hyperiidean amphipod superfamily Vibilioidea Bowman and Gruner, 1973 (Crustacea: Amphipoda: Hyperiidea). *Zootaxa* 280: 1–104.

Zeidler, W. 2004. A review of the hyperiidean amphipod superfamily Lycaeopsoidea Bowman & Gruner, 1973 (Crustacea: Amphipoda: Hyperiidea). *Zootaxa* 520: 1–18.

Zeidler, W. 2004. A review of the families and genera of the hyperiidean amphipod superfamily Phronimoidea Bowman & Gruner, 1973 (Crustacea: Amphipoda: Hyperiidea). *Zootaxa* 567: 1–66.

Zeidler, W. 2006. A review of the hyperiidean amphipod superfamily Archaeoscinoidea Vinogradov, Volkov & Semenova, 1982 (Crustacea: Amphipoda: Hyperiidea). *Zootaxa* 1125: 1–37.

Barnacles (Subphylum Crustacea, Class Cirripedia)

Buckeridge, J.S., and Newman, W.A. 2006. A revision of the Iblidae and the stalked barnacles (Crustacea: Cirripedia: Thoracica), including new ordinal, familial and generic taxa, and two new species from New Zealand and Tasmanian waters. *Zootaxa* 1136: 1–38.

Jones, D.S. 1987. A key to the common sessile barnacle species in the Swan-Canning river estuary, Western Australia. *Curtain University Environmental Studies Group Report* 1: 153–162.

Jones, D.S. 1990. A guide to the shallow-water barnacles (Cirripedia: Lepadomorpha, Balanomorpha) of the Shark Bay area, Western Australia. Pp. 209–229 in: Berry, P.B., Bradshaw, S.D., and Wilson, B.R. (eds) *Research in Shark Bay: Report of the France-Australe Bicentenary Expedition Committee.* Western Australian Museum: Perth.

Jones, D.S. 1990. The shallow-water barnacles (Cirripedia: Lepadomorpha, Balanomorpha) of southern Western Australia. Pp. 333–437 in: Wells, F.E., Walker, D.I., Kirkman, H., and Lethbridge, R. (eds) *Proceedings of the Third International Marine Biological Workshop: The marine flora and fauna of Albany, Western Australia.* 1st edn. Western Australian Museum: Perth.

Jones, D.S. 1991. A history of the discovery and description of Australian barnacles (Cirripedia: Thoracica), including a bibliography of reference works. *Archives of Natural History* 18: 149–178.

Jones, D.S. 1992. A review of Australian fouling barnacles. *Asian Marine Biology* 9: 89–100.

Jones, D.S. 1992. Scalpellid barnacles (Cirripedia: Thoracica) from the northestern and central eastern Australian continental shelf and slope. *Memoirs of the Queensland Museum* 32: 145–178.

Jones, D.S. 1993. The barnacles of Rottnest Island, Western Australia, with descriptions of two new species. Pp. 113–133 in: Wells, F.E., Walker, D.I., Kirkman, H., and Lethbridge, R. (eds) *Proceedings of the Fifth International Marine Biological Workshop: The marine flora and fauna of Rottnest Island, Western Australia.* Western Australian Museum: Perth.

Jones, D.S. 1998. New genus and species of Calanticidae (Cirripedia, Thoracica, Scalpellomorpha) from Australian waters. *Zoosystema* 20: 239–253.

Jones, D.S., Anderson, J.T., and Anderson, D.T. 1990. Checklist of the Australian Cirripedia. *Technical Reports of the Australian Museum* 3: 1–28.

Underwood, A.J. 1977. *Barnacles. A guide based on the barnacles found on the New South Wales coast.* Reed Education: Sydney. 32pp.

Molluscs (Phylum Mollusca)–general

Beesley, P.L., Ross, G.J.B., and Wells, A. (eds) 1998. *Mollusca: The Southern Synthesis. Fauna of Australia Vol. 5.* CSIRO Publishing: Melbourne. Part A xvi 563pp.; Part B viii 565–1234pp.

Ludbrook, N.H., and Gowlett-Holmes, K. 1989. Chitons, Gastropods and Bivalves (Shipworms, Family Teredinidae by J.Marshall-Ibrahim). Pp. 504–724 in: Shepherd, S.A., and Thomas, I.M. (eds) *Marine Invertebrates of Southern Australia. Part II.* South Australian Government Printing Division: Adelaide.

Macpherson, J.H., and Gabriel, C.H. 1962. *Marine Molluscs of Victoria.* Melbourne University Press: Melbourne.

Ponder, W.F., Clark, S.A., and Dallwitz, M.J. 2000. *Freshwater and Estuarine Molluscs. An interactive, illustrated key for New South Wales. [CD-ROM].* CSIRO Publishing: Melbourne.

Rudman, W.B. 1997. *Malacological Society of Australasia.* http://www.amonline.net.au/invertebrates/mal/malsoc/index.htm

Smith, B.J., Black, J.H., and Shepherd, S.A. 1989. Molluscan egg masses. Pp. 841–891 in: Shepherd, S.A., and Thomas, I.M. (eds) *Marine Invertebrates of Southern Australia. Part II.* South Australian Government Printing Division: Adelaide.

Wilson, B.R., and Gillett, K. 1971. *Australian Shells.* A.H. & A.W. Reed: Sydney.

Chitons (Phylum Mollusca, Class Polyplacophora)

Gowlett-Holmes, K.L. 1990. A review of the endemic Australian chiton genus *Bassethullia* Pilsbry, 1928 (Mollusca: Polyplacophora: Acanthochitonidae). *Journal of the Malacological Society of Australia* 11: 9–28.

Gowlett–Holmes, K.L. 1991. Redefinition of the genus *Notoplax* H. Adams, 1861, and recognition of the monotypic New Zealand genus *Pseudotonicia* Ashby, 1928 (Mollusca: Polyplacophora: Acanthochitonidae). *Journal of the Malacological Society of Australia* 12: 77–88.

Kaas, P., and Van Belle, R.A. 1985. *Monograph of Living Chitons (Mollusca: Polyplacophora). Vol. 1. Order Neoloricata: Lepidopleurina.* E.J. Brill/Dr W. Backhuys: Leiden. 240pp.

Kaas, P., and Van Belle, R.A. 1985. *Monograph of Living Chitons (Mollusca: Polyplacophora). Vol. 2. Suborder Ischnochitonina. Ischnochitonidae: Schizoplacinae, Callochitoninae & Lepidochitoninae.* E.J. Brill/Dr W. Backhuys: Leiden. 298pp.

Kaas, P., and Van Belle, R.A. 1987. *Monograph of Living Chitons (Mollusca: Polyplacophora). Vol. 3. Suborder Ischnochitonina. Ischnochitonidae: Chaetopleurinae & Ischnochitoninae (pars). Additions to Vols 1 & 2.* E.J. Brill/Dr W. Backhuys: Leiden. 302pp.

Kaas, P., and Van Belle, R.A. 1990. *Monograph of Living Chitons (Mollusca: Polyplacophora). Vol. 4. Suborder Ischnochitonina. Ischnochitonidae: Ischnochitoninae (continued). Additions to Vols 1, 2 and 3.* E.J. Brill/Dr W. Backhuys: Leiden. 298pp.

Kaas, P., and Van Belle, R.A. 1994. *Monograph of Living Chitons (Mollusca: Polyplacophora).*

Vol. 5. Suborder Ischnochitonina. Ischnochitonidae: Ischnochitoninae (concluded); Callistoplacinae; Mopaliidae. Additions to Vols 1–4. E.J. Brill/Dr W. Backhuys: Leiden. 402pp.

Ludbrook, N.H., and Gowlett–Holmes, K. 1989. Chitons, Gastropods and Bivalves (Shipworms, Family Teredinidae by J.Marshall-Ibrahim). Pp. 504–724 in: Shepherd, S.A., and Thomas, I.M. (eds) *Marine Invertebrates of Southern Australia. Part II.* South Australian Government Printing Division: Adelaide.

Mussels, clams (Phylum Mollusca, Class Bivalvia)

Lamprell, K., and Healy, J. 1992. *Bivalves of Australia Volume 1.* Crawford House Press: Bathurst, NSW. xii + 182pp.

Lamprell, K., and Healy, J. 1998. *Bivalves of Australia Volume 2.* Backhuys: Leiden. 288pp.

Middelfart, P. 2002. Revision of the Australian Cuninae *sensu lato* (Bivalvia: Carditoidea: Condylocardiidae). *Zootaxa* 112: 1–124.

Tusk shells (Phylum Mollusca, Class Scaphopoda)

Lamprell, K., and Healy, J.M. 1998. A revision of the Scaphopoda from Australian waters (Mollusca). *Records of the Australian Museum Supplement* 24: 1–189.

Lamprell, K., and Healy, J.M. 2001. Scaphopoda. Pp. 85–128 in: Wells, A., and Houston, W.W.K. (eds) *Zoological Catalogue of Australia. Vol. 17.2. Mollusca: Aplacophora, Polyplacophora, Scaphopoda, Cephalopoda.* CSIRO Publishing: Melbourne.

Snails, slugs (Phylum Mollusca, Class Gastropoda)

Bouchet, P., and Rocroi, J.-P. (eds). 2005. Classification and nomenclator of gastropod families. *Malacologia* 47: 1–397.

Burn, R. 1989. Opisthobranchs (Subclass Opisthobranchia). Pp. 725–788 in: Shepherd, S.A., and Thomas, I.M. (eds) *Marine Invertebrates of Southern Australia. Part II.* South Australian Government Printing Division: Adelaide.

Burn, R., Museum Victoria, and ReefWatch Victoria. 2006. *Nudibranchs and relatives–a photographic atlas of the opisthobranchs of the Bass Strait Region.* http://researchdata.museum.vic.gov.au/marine/nudi_home.htm

Coleman, N. 2001. *1001 Nudibranchs–Catalogue of Indo-Pacific Sea Slugs.* Neville Coleman's Underwater Geographic: Springwood, Queensland.

Coovert, G.A., and Coovert , H.K. 1995. Revision of the supraspecific classification of marginelliform gastropods. *The Nautilus* 109: 43–110.

Darragh, T.A. 2002. A revision of the Australian genus *Umbilia* (Gastropoda: Cypraeidae). *Memoirs of Museum Victoria* 59: 355–392

Geiger, D.L., and Jansen, P. 2004. Revision of the Australian species of Anatomidae (Mollusca: Gastropoda: Vetigastropoda). *Zootaxa* 415: 1–35.

Geiger, D.L., and Jansen, P. 2004. New species of Australian Scissurellidae (Mollusca: Gastropoda: Vetigastropoda) with remarks on Australian and Indo-Malayan species. *Zootaxa* 714: 1–72.

Ponder, W.F. 1999. *Calopia* (Calopiidae), a new genus and Family of estuarine gastropods

(Caenogastropoda: Rissooidea) from Australia. *Molluscan Research* 20: 17–60.

Ponder, W.F. 2003. Monograph of the Australian Bithyniidae (Caenogastropoda: Rissooidea). Zootaxa 230: 1–126.

Rudman, W.B. 2006. *Seaslug Forum*. http://www.seaslugforum.net/

Smith, B.J., and Kershaw, R.C. 1979. *Field Guide to the Non-Marine Molluscs of South Eastern Australia*. Australian National University Press: Canberra. 285pp. [also treats some mollusc groups occupying the littoral fringe, mangrove, salt marsh & estuarine habitats]

Wilson, B.R. 1993. *Australian Marine Shells. Prosobranchs Gastropods, Part One*. Odyssey Publishing: Kallaroo, Western Australia. 408pp.

Wilson, B.R. 1994. *Australian Marine Shells. Prosobranchs Gastropods, Part Two*. Odyssey Publishing: Kallaroo, Western Australia. 370pp.

Aplacophorans (obscure worm-like molluscs, Phylum Mollusca, Class Aplacophora)

Scheltema, A.H. 1989. Australian aplacophoran molluscs. I. Chaetodermamorpha from Bass Strait and the continental slope off south-eastern Australia. *Records of the Australian Museum* 41: 43–62.

Scheltema, A.H. 1999. Two Solenogaster Molluscs, *Ocheyoherpia trachia* n.sp. from Macquarie Island and *Tegulaherpia tasmanica* Salvini-Plawen from Bass Strait (Aplacophora: Neomeniomorpha). *Records of the Australian Museum* 51: 23–31.

Monoplacophorans (obscure deepwater limpet-like molluscs, Phylum Mollusca, Class Monoplacophora)

Giribet, G., Okusu, A., Lindgren, A.R., Huff, S.W., Schrodl, M., and Nishiguchi, M.K. 2006. Evidence for a clade composed of molluscs with serially repeated structures: Monoplacophorans are related to chitons. *Proceedings of the National Academy of Sciences, Philadelphia* 103: 7723–7728.

Squid, cuttlefish, octopuses (Phylum Mollusca, Class Cephalopoda)

Lu, C.C. 1998. A Synopsis of Sepiidae in Australian waters (Cephalopoda: Sepioidea). Systematics & biogeography of cephalopods Vol. 1. *Smithsonian Contributions to Science* 586: 159–190.

Lu, C. C. (2001). Cephalopoda. *Zoological Catalogue of Australia* A. Wells and Houston, W. W. K. Melbourne, CSIRO Publishing. 17.2: 129-308.

Lu, C.C. and M.C. Dunning (1998). Subclass Coleoidea. pp.499-563. Beesley, P.L., Ross, G.J.B. & A. Wells. (eds) *Mollusca: The Southern Synthesis. Fauna of Australia*. Vol. 5. CSIRO Publishing: Melbourne. Part A, xvi 563pp.

Mangold, K., M.R. Clarke and C.F.E. Roper (1998). Cephalopod Introduction. pp. 451-484. Beesley, P.L., Ross, G.J.B. & A. Wells. (eds) *Mollusca: The Southern Synthesis. Fauna of Australia. Vol.* 5. CSIRO Publishing: Melbourne. Part A, xvi 563pp.

Norman, M.D. 2000. *Cephalopods, a World Guide*. Conchbooks: Hackenheim. 320pp.

Norman, M.D., and Reid, A. 2000. *A Guide to Squid, Cuttlefish and Octopuses of Australasia*. CSIRO Publishing: Melbourne. 96pp.

Norman, M.D. 2002. *Species Bank: Cephalopods.* www.environment.gov.au/biodiversity/abrs/online-resources/species-bank/index.html

Norman, M.D., and Finn, J.K. 2006. *Argosearch.* www.argosearch.org.au

Tree of Life website:
Cephalopoda: http://tolweb.org/tree?group=Cephalopoda&contgroup=Mollusca

Lace coral, bryozoans (Phylum Bryozoa)

Bock, P.E. 1982. Bryozoans (Phylum Bryozoa or Ectoprocta). Pp. 319–394 in: Shepherd, S.A., and Thomas, I.M. (eds) *Marine Invertebrates of Southern Australia Part 1.* Handbooks Committee of the South Australian Government: Adelaide.

Bock, P.E., and Cook, P.L. 2004. A review of Australian Conescharellinidae (Bryozoa: Cheilostomata). *Memoirs of Museum Victoria* 61: 135–182.

Bock, P.E. 2006. *Recent and Fossil Bryozoa.* http://www.civgeo.rmit.edu.au/bryozoa/default.html

Gordon, D.P. 2004. *New Zealand Bryozoan Biodiversity Database. Web publication.* http://nzbbd.niwa.co.nz/

Schmidt, R. 2006. *International Bryozoology Association.* http://www.nhm.ac.uk/hosted_sites/iba/

Waggoner, B., and Collins, A.G. 1999. *Introduction to the Bryozoa.* Museum of Paleontology, University of California: Berkeley.

Lamp shells (Phylum Brachiopoda)

Dawson, E. 1990. The systematics and biogeography of the living brachiopoda of New Zealand. Pp. 431–437 in: Lee, D., and Campbell, D.J. (eds) *Brachiopods through Time.* A.A. Balkema: Rotterdam.

Richardson, J.R. 1997. Brachiopods (Phylum Brachiopoda). Pp. 999–1027 in: Shepherd, S.A., and Davies, M. (eds) *Marine Invertebrates of Southern Australia Part III.* Handbooks Committee of the South Australian Government: Adelaide.

Echinoderms–general (Phylum Echinodermata)

Clark, A.M., and Rowe, F.W.E. 1971. *Monograph of Shallow-water Indo-west Pacific Echinoderms.* British Museum: London. 238pp.

Marsh, L.M. 1990. Shallow water echinoderms of the Albany region, south-western Australia. Pp. 439–482 in: Wells, F.E., Walker, D.I., Kirkman, H., and Lethbridge, R. (eds) *Proceedings of the Third International Marine Biological Workshop: The marine flora and fauna of Albany, Western Australia.* Western Australian Museum: Perth.

Marsh, L.M., and Pawson, D.L. 1999. Echinoderms of Rottnest Island. Pp. 279–304 in: Walker, I., and Wells, F.E. (eds) *The Seagrass Flora and Fauna of Rottnest Island, Western Australia, Perth.* Western Australian Museum: Perth.

Feather stars (Phylum Echinodermata, Class Crinoidea)

Shepherd, S.A., Marshall, J., and Rowe, F.W.E. 1982. Feather-stars (Class Crinoidea). Pp. 396–399 in: Shepherd, S.A., and Thomas, I.M. (eds) *Marine Invertebrates of Southern Australia Part 1*. Handbooks Committee of the South Australian Government: Adelaide.

Sea stars (Phylum Echinodermata, Class Asteroidea)

Marsh, L.M. 1990. A revision of the echinoderm genus *Bunaster* (Asteroidea: Ophidiasteridae). *Records of the Western Australian Museum* 15: 419–433.

O'Loughlin, P.M. 2002. New genus and species of southern Australian and Pacific Asterinidae (Echinodermata, Asteroidea). *Memoirs of Museum Victoria* 59: 277–296.

O'Loughlin, P.M., and O'Hara, T.D. 1990. A review of the genus *Smilasterias* (Echinodermata, Asteroidea), with descriptions of two new species from south-eastern Australia, one a gastric brooder, and a new species from Macquarie Island. *Memoirs of Museum Victoria* 50: 307–323.

O'Loughlin, P.M., and Waters, J.M. 2004. A molecular and morphological revision of genera of Asterinidae (Echinodermata: Asteroidea). *Memoirs of Museum Victoria* 61: 1–40.

O'Loughlin, P.M., Waters, J.M., and Roy, M.S. 2003. A molecular and morphological review of the asterinid, *Patiriella gunnii* (Gray) (Echinodermata: Asteroidea). *Memoirs of Museum Victoria* 60: 181–195.

Zeidler, W., and Shepherd, S.A. 1982. Sea-stars (Class Asteroidea). Pp. 400–417 in: Shepherd, S.A., and Thomas, I.M. (eds) *Marine Invertebrates of Southern Australia Part I*. Handbooks Committee of the South Australian Government: Adelaide.

Brittle stars (Phylum Echinodermata, Class Ophiuroidea)

Baker, A.N. 1982. Brittle-stars (Class Ophiuiroidea). Pp. 418–436 in: Shepherd, S.A., and Thomas, I.M. (eds) *Marine Invertebrates of Southern Australia Part I*. Handbooks Committee of the South Australian Government: Adelaide.

O'Hara, T.D. 1990. New records of Ophiuridae, Ophiacanthidae and Ophiocomidae (Echinodermata: Ophiuroidea) from south-eastern Australia. *Memoirs of Museum Victoria* 50: 287–305.

Sea urchins (Phylum Echinodermata, Class Echinoidea)

Baker, A.N. 1982. Sea-urchins (Class Echinoidea). Pp. 437–453 in: Shepherd, S.A., and Thomas, I.M. (eds) *Marine Invertebrates of Southern Australia Part I*. Handbooks Committee of the South Australian Government: Adelaide.

Miskelly, A. 2003. Sea Urchins of Australia and the Indo-Pacific. P. 118pp. in. Capricornia Publications: Linfield, NSW, Australia.

Sea cucumbers (Phylum Echinodermata, Class Holothuroidea)

O'Loughlin, P.M. 1998. A review of the holothurian family Gephyrothuriidae. Pp. 493–498 in: Mooi, R., and Telford, M. (eds) *Echinoderms: San Francisco; Proceedings of the ninth*

International Echinoderm Conference San Francisco/California/USA/5–9 August 1996. Balkema: Rotterdam.

O'Loughlin, P.M. 2000. A review of the cucumariid genus *Psolidiella* Mortensen (Echinodermata, Holothuroidea). *Memoirs of Museum Victoria* 58: 25–37.

O'Loughlin, P.M. 2002. Report on selected species of BANZARE and ANARE Holothuroidea, with reviews of *Meseres* Ludwig and *Heterocucumis* Panning (Echinodermata). *Memoirs of Museum Victoria* 59: 297–326.

O'Loughlin, P.M., and Alcock, N. 2000. The New Zealand Cucumariidae (Echinodermata, Holothuroidea). *Memoirs of Museum Victoria* 58: 1–24.

O'Loughlin, P.M., and O'Hara, T.D. 1993. New cucumariid holothurians (Echinodermata) from southern Australia, including two brooding and one fissiparous species. *Memoirs of the Museum of Victoria* 53: 227–266.

Rowe, F.W.E. 1982. Sea-cucumbers (Class Holothuroidea). Pp. 454–476 in: Shepherd, S.A., and Thomas, I.M. (eds) *Marine Invertebrates of Southern Australia Part I*. Handbooks Committee of the South Australian Government: Adelaide.

Acorn worms and pterobranchs (Phylum Hemichordata)

Burdon-Jones, C. 1998. Hemichordata. Pp. 1–50 in: Wells, A., and Houston, W.W.K. (eds) *Zoological Catalogue of Australia*. CSIRO Publishing: Melbourne.

Cameron, C.B. 2006. Hemichordata images; Checklist of hemichordate species; Taxonomic key to the enteropneusta. http://cluster3.biosci.utexas.edu/faculty/cameronc/CBC.htm

Shepherd, S.A. 1997. Acorn worms and a Pterobranch (Phylum Hemichordata). Pp. 1028–1039 in: Shepherd, S.A., and Davies, M. (eds) *Marine Invertebrates of Southern Australia Part III*. Handbooks Committee of the South Australian Government: Adelaide.

Arrow worms (Phylum Chaetognatha)

O'Sullivan, D., and Hosie, G. 1985. A general guide to the metazoan zooplankton groups of the Southern Ocean. *ANARE Research Notes* 30: 1–59.

Thomson, J.M. 1947. The Chaetognaths of south-eastern Australia. *Council for Scientific and Industrial Research Bulletin, Australia* 222: 1–43.

Lancelets (Phylum Chordata, Subphylum Cephalochordata)

Last, P.R., Scott, E.O.G., and Talbot, F.H. 1983. *Fishes of Tasmania*. Tasmanian Fisheries Development Authority: Hobart. vii + 563pp.

Richardson, B.J., and McKenzie, A.M. 1994. Taxonomy and distribution of Australian Cephalochordates (Chordata: Cephalochordata). *Invertebrate Taxonomy* 8: 1443–14459.

Sea squirts (Phylum Chordata, Subphylum Urochordata, Class Ascidiacea)

Kott, P. 1985. The Australian Ascidiacea Pt 1, Phlebobranchia and Stolidobranchia. *Memoirs of the Queensland Museum* 23.

Kott, P. 1990. The Australian Ascidiacea Pt 2, Aplousobranchia (1). *Memoirs of the Queensland Museum* 29.

Kott, P. 1990. The Australian Ascidiacea, Phlebobranchia and Stolidobranchia, supplement. *Memoirs of the Queensland Museum* 29.

Kott, P. 1992. The Australian Ascidiacea Pt 3, Aplousobranchia (2). *Memoirs of the Queensland Museum* 32.

Kott, P. 1992. The Australian Ascidiacea, supplement 2. *Memoirs of the Queensland Museum* 32.

Salps (Phylum Chordata, Subphylum Urochordata, Class Thaliacea)

O'Sullivan, D. 1982. A guide to the pelagic tunicates of the Southern Ocean and adjacent waters. *ANARE Research Notes* 8: 1–98.

Thompson, H. 1948. Pelagic tunicates of Australia. *Handbook, Council for Scientific and Industrial Research (Australia)*: 1–96.

Vertebrates–Hagfish, lampreys, sharks, fish, birds, reptiles, mammals (Phylum Chordata, Subphylum Vertebrata)

Bryden, M., Marsh, H., and Shaughnessy, P. 1998. *Dugongs, Whales, Dolphins and Seals. A guide to the sea mammals of Australasia.* Allen and Unwin: St Leonards, New South Wales. 17pp.

Glasby, C.J., Ross, G.J.B., and Beesley, P.L. (eds) 1993. *Amphibia and Reptilia. Fauna of Australia: Vol 2A.* CSIRO Publishing and Australian Biological Resources Study: Melbourne. xii + 465pp.

Gomon, M.F., Glover, J.C.M., and Kuiter, R.H. 1994. *The Fishes of Australia's South Coast.* The Flora and Fauna of South Australia Handbooks Committee: Adelaide. 992pp.

Heatwole, H. 1999. *Sea Snakes.* University of New South Wales Press: Sydney. 148pp.

Kuiter, R.H. 2000. *Coastal Fishes of South-eastern Australia.* Crawford House Press: Bathurst, New South Wales. xxxii + 437pp.

Lane, B.A., and Davies, J.N. 1987. *Shorebirds in Australia.* Nelson: Melbourne. 187pp.

Last, P.R., Scott, E.O.G., and Talbot, F.H. 1983. *Fishes of Tasmania.* Tasmanian Fisheries Development Authority: Hobart. vii + 563pp.

Menkhorst, P.W. 1995. *Mammals of Victoria: distribution, ecology and conservation.* Oxford University Press: Melbourne. viii + 360pp.

Menkhorst, P.W., and Knight, F. 2004. *A Field Guide to the Mammals of Australia.* 2nd edn. Oxford University Press: Melbourne. 280pp.

Simpson, K.G. 1972. *Birds in Bass Strait.* A.H. & A.W.Reed for Broken Hill Propietary Co Ltd.: Sydney. 112pp.

Slater, P., Slater, R., and Slater, P. 1989. *The Slater field guide to Australian birds.* Weldon: Sydney. 34pp.

Walton, D.W., and Richardson, B.J. (eds) 1989. *Mammalia. Fauna of Australia Volume 1B.* CSIRO Publishing and Australian Biological Resources Study: Melbourne. 827pp.

Wilson, S., and Swan, G. 2003. *A Complete Guide to Reptiles of Australia.* New Holland. 480pp.

INDEX OF SCIENTIFIC NAMES

Page references in **bold** indicate an image of the subject

INDEX OF COMMON NAMES

Page references in **bold** indicate an image of the subject

Other titles in the series

Crabs, hermit crabs and allies
Shrimps, prawns and lobsters
Barnacles

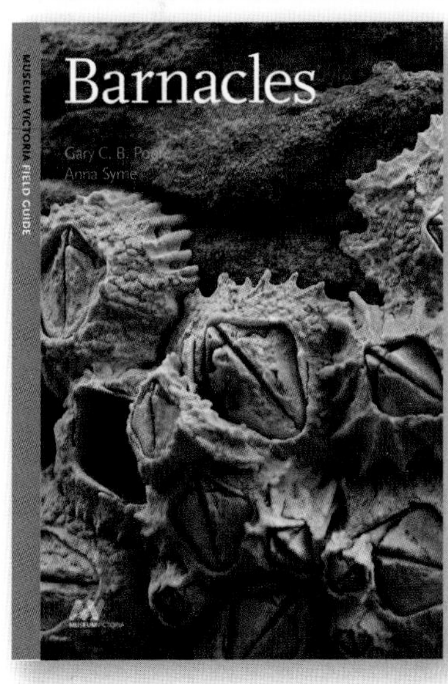